Shavers Fork
NATIONAL FOREST
Cheat Mountain
DURBIN
Bartow
Spruce Knob
Boyer Hill
Mennonite Church
Trent's
General Store
ARBOVALE
Observatory
Elementary-Middle School
Public Library
GREEN BANK
10-mile radius
N
100 miles
PITTSBURGH
PENNSYLVANIA
COLUMBUS
OHIO
MARYLAND
WEST VIRGINIA
WASHINGTON, D.C.
POCAHONTAS
COUNTY
Sugar Grove
Station
Green Bank
Observatory
CHARLESTON
KENTUCKY
CHARLOTTESVILLE
WHITE SULPHUR
SPRINGS
National Radio
Quiet Zone (NRQZ)
RICHMOND
VIRGINIA

THE QUIET ZONE

THE QUIET ZONE

UNRAVELING THE MYSTERY OF A TOWN SUSPENDED IN SILENCE

STEPHEN KURCZY

DEY ST.
An Imprint of WILLIAM MORROW

HarperCollins books may be purchased for educational, business, or sales promotional use. For information, please email the Special Markets Department at SPsales@harpercollins.com.

FIRST EDITION

Designed by Michelle Crowe
Endpaper map design by Mike Hall
Background image by detchana wangkheeree/Shutterstock, Inc.

Library of Congress Cataloging-in-Publication Data has been applied for.

ISBN 978-0-06-294549-5

21 22 23 24 25 LSC 10 9 8 7 6 5 4 3 2 1

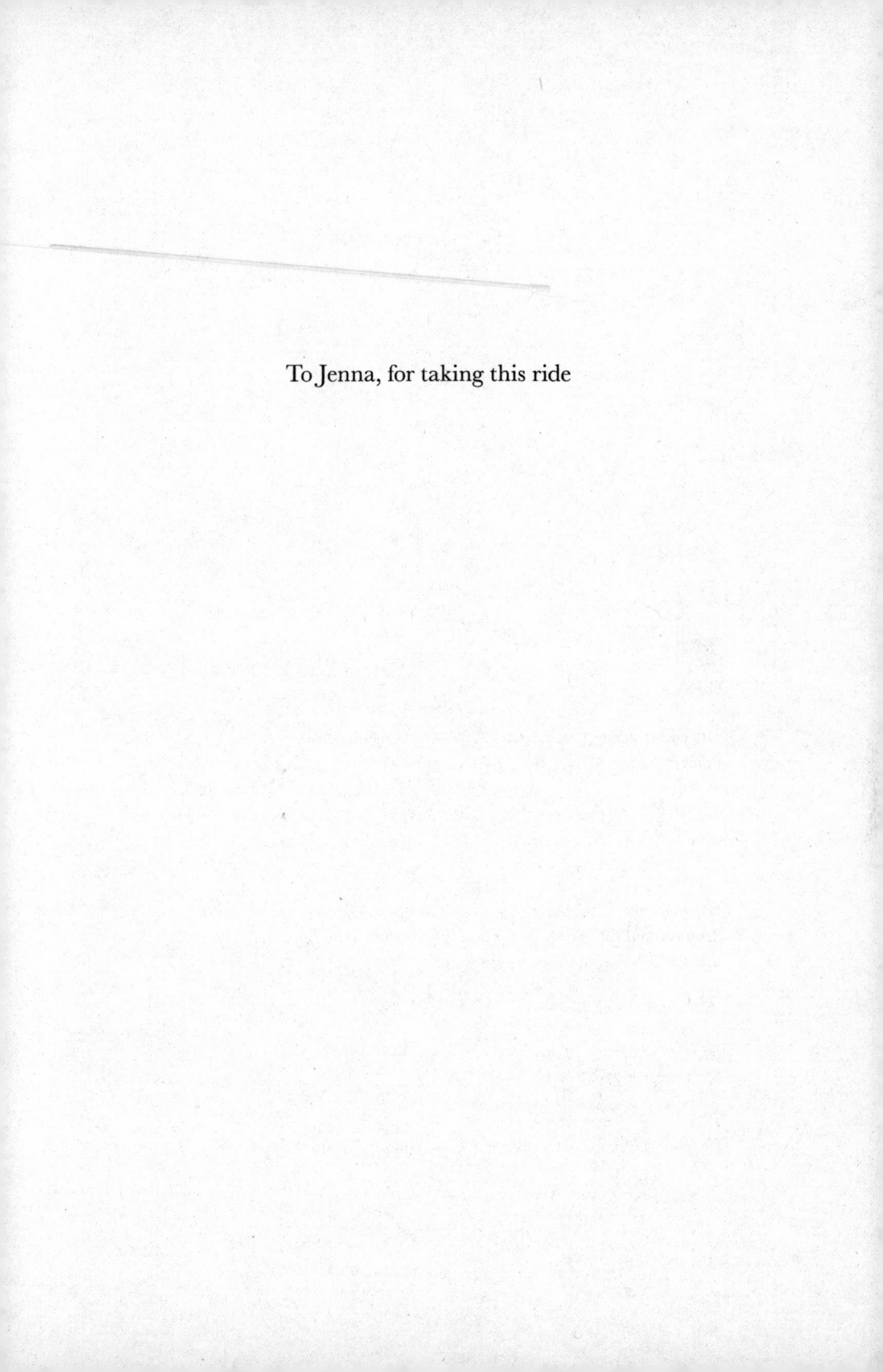

To Jenna, for taking this ride

CONTENTS

PART THREE:
QUIET END?

PROLOGUE

"To Anyone Who Will Listen"

DEEP IN THE MOUNTAINS OF APPALACHIA, on a cold January afternoon, I scanned the trees in search of a simple wooden cross, the kind that might mark the site of a roadside car accident and warn passersby, *Don't let this happen to you*. I suddenly felt I was being watched. I heard an engine rev. I looked back toward the gravel parking lot of a Mennonite church several hundred feet away and watched a truck loop around my car and peel off.

I turned back to the shadowy forest, searching among the fallen leaves for the cross. I was told it would be little more than two tree branches tied together with a vine and propped upright, the place where a woman had killed herself. Her death was not reported in the local newspaper, but it was whispered about in the community. *She was allergic to WiFi,* people were saying.

Circling among the spindly oaks and maples, all I could find were old bottles, empty beer cans, and animal bones—roadkill remnants, perhaps appropriate for a county where the biggest tourism event was the annual Roadkill Cook-Off. I was a few miles outside of Green Bank, West Virginia, a remote community with a claim to being "the quietest town in America"—which was what had first drawn me here years earlier. In Green Bank, the nation's oldest federal radio astronomy observatory operated a collection of giant, dish-shaped telescopes that measured the invisible energy waves raining down on Earth from the heavens. To detect those faint radio waves, the observatory demanded quiet from the surrounding community. I stood near the center of the National Radio Quiet Zone. It was a place where cellphone signals, WiFi, and other electronic noise were tightly monitored and restricted.

The quiet had attracted a number of groups over the decades, the latest being people who sought refuge from our increasingly digital, electrified world. These people described feeling ill when exposed to iPhones and smart meters, refrigerators and microwaves. In essence, they felt allergic to modern life. And many felt they had nowhere to go but Green Bank. They worshipped the quiet here, walking barefoot beneath the massive radio telescopes, one of which was taller than the Statue of Liberty. It was like a Statue of Quiet, marking this as a Holy Land of silence. Or so I had thought.

Now there was this other marker, a cross in the woods. The woman who killed herself had been smart, driven, and compassionate, a graduate of Vassar College and Harvard Business School. She had worked on Wall Street, then dropped out of the corporate world to advocate for animal rights and care for disabled people in the Charlottesville area of Virginia, until she felt debilitated by the onset of a new disease called "electromagnetic hypersensitivity" (EHS). In September 2018, she came to Green Bank.

"The world is a wondrous place, full of beauty, mystery, miracles, and love," the woman wrote in a final, handwritten letter found in her car and addressed "to anyone who will listen." "The world is also full of perils, known and unknown, visible and invisible. I am writing now about a mostly unknown, invisible peril—electromagnetic frequencies, or EMFs . . . Please do not let our children grow up in an inescapable sea of invisible, insidious waves."

After writing the letter, the woman had parked outside the Mennonite church, walked into the woods where I stood, and shot herself in the head with a .38 Smith & Wesson revolver. It was likely the first time she ever used a gun, according to her family and friends. At least one neighbor heard the shot but thought nothing of it, gunshots being common in an area known for its hunting. After three days, a road crew found the body. Authorities, family, and friends began piecing together the woman's final days and months.

What had driven such a successful, empathetic woman to purchase a gun and end her life one evening? She appeared to have sacrificed herself in an attempt to call attention to our very human need for quiet and raise the alarm about its endangered state. But who would listen to her?

PART ONE

QUIET SEARCH

> They found rich, good pasture, and the land was spacious, peaceful, and quiet.
>
> —1 CHRONICLES 4:40

CHAPTER ONE

"Over the Mountain"

DEER CREEK VALLEY WAS DARKENING, the surrounding ring of forested peaks fading into the clouds. Jenna and I had been on the road all day, driving through the Allegheny range of West Virginia and the seemingly abandoned towns of Pocahontas County. Through stretches of thick forest and rolling mountains, the road wound uphill for miles and then careened downward at deadly steep grades. Big-eyed cows stared as we drove by their pastures. Occasionally, we'd see a gas station and think, *Oh, there* is *civilization here.*

Just outside the town of Green Bank, I parked alongside a clapboard church—we'd seen more churches than people that day. I got out of the car and crunched over the glazed snow to the Chestnut Ridge Country Inn next door. I knocked—no answer. I turned the knob—the door was locked. On the porch of the colonial home was a chalkboard that read "Welcome Jesse and Jennifer," dated February 2017. But we were not Jesse and Jennifer, and it was March.

I glanced at Jenna, who appeared increasingly anxious. We had no cell service, no WiFi, and nowhere to sleep. Her iPhone searched in vain for a signal, its status wheel spinning like a compass inside the Bermuda Triangle. A few miles away, a platter-size road sign had vaguely explained the reason for our disconnection: "You Are

Now Entering the West Virginia Radio Quiet Zone." A cat leaped onto the porch and nuzzled my sneaker, oblivious to my unease. I felt like a child in the silent woods, spooked by how loud quiet can be. Feeling untethered and lost, we were struggling to answer a basic question: *Where will we sleep tonight?*

We climbed back into the car and drove five miles to Henry's Quick Stop, a gas station that also served as a grocery store, restaurant, and ice-cream parlor, selling everything from scratch-offs to gun ammo in Green Bank, which had an estimated population of 250. It was our third time at Henry's that afternoon. We'd first pulled in for gas before trying to check into the nearby Boyer Motel, a manila-colored structure that reminded me of the Bates Motel from *Psycho;* it was closed, perhaps for the best. We'd then returned to Henry's and gotten directions to the Chestnut Ridge Country Inn. Now back at Henry's, I asked the bearded attendant where else we might spend the night. He shrugged. Jenna opened a tourist brochure and saw a listing for lodging in Durbin, about ten miles north on the sole road that cut through town.

"Could I call from here?" I asked.

"Go on ahead," the attendant said, gesturing to a landline beside the register. "Store closes in fifteen minutes. Streets roll up at seven."

He handed me a heavy phone book. (When had I last used a phone book?) Its thin pages held the names and numbers of Pocahontas County's 8,200 residents—about one-tenth the population of the New York City neighborhood where we lived. I flipped through and found the number for a place called Station 2.

"We've got space," a woman said over the phone. "But we're five minutes to closing so you'd better hurry up."

As we raced through town with a pepperoni pizza from Henry's, we passed the area's quiet authority: the Green Bank Observatory, founded in 1956 by the National Science Foundation (NSF). We could see a handful of radio telescopes poking above the trees, the largest a 485-foot-tall tangle of white beams holding a giant dish the size of two football fields. It looked like a washbasin for Godzilla. The telescopes sat at the bottom of a four-mile-long valley surrounded by mountains up to 4,800 feet tall, which created a natural barrier against the outside world's noise. Operating any electrical equipment within ten miles of here was illegal if it caused interference to the telescopes, punishable by a state fine of fifty dollars per day. Surrounding that ten-mile radius, a thirteen-thousand-square-mile National Radio Quiet Zone—an area larger than the combined landmass of Connecticut and Massachusetts—further limited cell service and all kinds of wireless communications systems. The restrictions were based on a simple premise: To listen, we have to hear. To unlock the mysteries of the universe, we have to be quiet.

The physical and bureaucratic barriers isolated an already remote area. In terms of the absence of man-made electronic noise, no other modern-day community was considered as quiet. A handful of other radio quiet zones existed worldwide, but they were in essentially uninhabited areas. Green Bank was a living, breathing community—though sparsely populated, to be sure. Three-fifths of the surrounding county of Pocahontas was state or federal forest. Its 941 square miles had a total of three traffic lights and three official towns. (Green Bank, as an unincorporated community, was not among them.) Residents shared one weekly newspaper, one high school, and a couple roadside telephone booths. The population density of about nine people per square mile was the lowest in West Virginia and one of the lowest anywhere east of the Mississippi River. Going to Walmart was a hundred-mile round trip that required traversing some of the Mountain State's tallest peaks. Outsiders were considered "flatlanders" or "come-heres." Locals were "mountain people." History crept forward in a place like this; many residents knew which great-great-grandparent settled the land and on which side their great-grandfather fought during the Civil War.

Earlier in the day, Jenna and I had stopped at a scenic overlook of the Monongahela National Forest, an expanse of rolling hills and layered mountain ridges covered in pine trees speckled white with snow. While West Virginia was known for its mining industry, this area of the state largely lacked coal, which had spared it from land-scarring strip-mining and mountaintop removal practices. The evergreen forests were thick with mountain laurel and, in warmer months, teeming with mushrooms, ramps, ginseng, goldenseal, and sassafras. The county was the source of eight major rivers flowing to the Atlantic and Gulf of Mexico. It was a land with evocative names like Stony Bottom, Clover Lick, Thorny Creek, Briery Knob, and Green Bank, with that last name holding an almost mythical allure as a place where the grass was greener and life was fuller. Four hours from Washington, D.C., Green Bank sounded to Jenna and me like a modern-day Walden that could free us from the exasperating demands of being always online and always reachable. Visiting was to be a respite from our digital lives.

A truck had parked beside us at the scenic overlook. An older man got out and waved. His wife, Patty, sat in the truck with their two dogs. We all started chatting. I asked the man, Les, if there was ever a moment when he'd wished for cell service. He launched into a half-hour tale about going hunting, slicing his hand to the bone as he gutted a deer, and then trudging miles to the nearest road while he bled through a makeshift tourniquet. "If I'd had a cellphone, I could have had a truck waiting for me," Les said. "But other than that time . . ." By this point I'd shoved my freezing hands deep inside my pockets, though the chill didn't seem to bother Les, who was chain-smoking Marlboros with gloveless hands; he'd smoked so many Marlboros over the years that he'd purchased his red Marlboro-branded jacket with points accumulated from the cigarette packs. No devices could interrupt our conversation, nobody could zone out on a smartphone. Before we parted, Les and Patty had invited us to their house for spaghetti and meatballs. (When had I last been invited to a stranger's home for dinner?)

We pulled into Durbin, a bygone logging town on the Greenbrier River. An old railway track was saddled with rusting boxcars. A row of boarded-up storefronts held our destination: Station 2, a combination hair salon, greasy spoon, and four-room motel. We parked in an empty lot. As we trudged up a set of wooden stairs to the entryway, a wiry man with translucent eyes that matched his pale complexion swung open the door and stared hard at me.

"We're closed," he grumbled.

"Thanks," I said, "but they're expecting us."

"No," he said. "We're *closed*."

"It's all right," I said, skirting around him, "we're sleeping here."

Shouldn't have said that, I thought.

IN THE FOYER OF STATION 2, a blond woman stood behind a cash register, which was perched on a glass cabinet filled with hunting knives and boxes of gun ammo. She explained that the restaurant was owned by the local fire chief, hence the decorative fire hose and thick bunker coat hanging on one wall. Charging us $77.28 for the night, she gave us a room key and led us through the kitchen and up a dark stairway. I mentioned that we'd tried to stay at the Boyer Motel, but it seemed abandoned.

"They don't even have WiFi," she said.

"There's WiFi here?" I asked, the excitement in my voice betraying my craving to get online.

"It's free under the name 'Station 2.'"

While we were still within the Quiet Zone and without cellphone reception, we were now far enough from Green Bank's telescopes for the legal use of WiFi, apparently. We were given a room with a double bed and flat-screen television. Before my bag hit the floor, I was logged on to the internet with my iPod. Soon my laptop was also connected, releasing a flood of emails and alerts that I'd missed over the previous twenty-four hours. The radio silence was broken. Jenna scrolled on her iPhone. We had teleported into separate worlds.

"We should check out that bar in town," Jenna said after a while, referencing a joint that we'd spotted on the drive into Durbin.

I didn't look up from my laptop.

"Let's go for one drink," she prodded—not that she needed a nightcap, just that she thought we should do more with our evening than stare at tiny screens.

We walked up the deserted street to Al's Upper Inn, the only establishment still open at the ungodly hour of 8:30 P.M. All conversations stopped as we entered. Jenna is Korean and I look like a nerdy white journalist, which is to say that we looked like outsiders. A half dozen people stared at us.

"We're not from around here," I said awkwardly.

"No kidding," someone replied. Chuckles.

We eased onto barstools and made small talk. I mentioned we were in town to visit the astronomy observatory. "Better get there before it closes," someone muttered, alluding to the facility's financial troubles. Several couples stared at a sports game on the wall-mounted TV. Two middle-aged men stood up to take a turn at a billiards table, one of them sporting a KKK tattoo on his biceps. He told me his name was J.R. and, unprompted, added that he hated the Puerto Rican migrants who were stealing local jobs. After his pool game, J.R. purchased six bottles of Budweiser to go before peeling away on a four-wheeler with a woman on back.

A stern-looking bartender hovered by the beer taps. I mentioned that I was fascinated with the local way of life, how the area felt like stepping back in time. The bartender rolled her eyes as if to say, *You don't know the half of it.* She told us of a saying, "Goin' over the mountain," which was when someone was heading out and would be unavailable by cellphone. We were way, *way* over the mountain.

For me, coming here was something of a pilgrimage. I hadn't owned a cellphone in nearly a decade, even as everyone around me increasingly did, from my elderly grandmother to my prepubescent niece and nephew. More than ever, I felt that I was in an ideological battle against a culture of constant connectivity, fighting the pressure to be like everyone else and get a smartphone. I had conceded to getting an iPod at some point over the years, and even with that pared-down device I sometimes felt as tech-addled as anyone, which was partly why I didn't want to take the next step of getting an iPhone. Was this remote area of West Virginia the last place where I could resist its influence? The last place where I could fit in without a smartphone?

IN A SENSE, my journey to the Quiet Zone began in 2009, when I got rid of my first and last cellphone. I had been living in Cambodia for two years, working as a reporter for the *Cambodia Daily* newspaper and traveling around the region to cover stories. My cellphone was so often at hand that it became an extension of myself. I slept with it. I ate with it. It was a social lifeline. It was also a source of anxiety. In need of a last-minute quote, desperate for a callback from a source, I would stare at the device, *willing* it to comply. I heard phantom rings and felt phantom vibrations. I was as dependent on my phone as a baby on a pacifier—a real condition, as the marketing professors Shiri Melumad and Michel Pham found in the 2017 research paper "Understanding the Psychology of Smartphone Usage: The Adult Pacifier Hypothesis." The day I left Cambodia, I dropped my flip phone in a garbage can. I wanted a break.

Back in the United States, I put off getting a replacement. It was a decision based on frugality—I was reluctant to sign a contract that would lock me into a payment plan. Weeks without a cellphone turned into months, then years. I worked for the *Christian Science Monitor* in Boston, then moved to New York City to report on finance, then relocated to Brazil as a foreign correspondent, all without a cellphone. I signed up for a free Google "phone number" that allowed me to make calls using my laptop. I used Skype. I got an iPod Touch for podcasts. In emergency situations, I borrowed others' cellphones. Once, on a 150-mile bicycle ride, I used a stranger's device to notify my family that I'd be arriving hours late and after dark, using it in the same way that people once utilized roadside pay phones, until they disappeared because everybody but me got a cellphone. I recognize that mobile devices can be useful. I just think they should be used sparingly and mostly in emergencies.

Family, friends, and colleagues began to question whether I was disconnected from the modern world or from reality. Employers grew irritated. "Get a cellphone and get on Facebook," an editor once told me. I declined both directives, but I agreed to at least open a Twitter account to "promote" our stories. My mother was frustrated that she couldn't keep tabs on me the way she did my smartphone-toting sisters. "I just worry about you," she'd say, in the way that mothers do. The more pressure I got, the more I dug in my heels. Why was the onus on me to change? After all, I was the normal person, by measure of how long humans had lived without cellphones. Wasn't I free to not have a cellphone?

I started to see it as a matter of personal liberty, a kind of Fourth Amendment fight for privacy and "the right to be let alone," as phrased by the Boston lawyers Samuel Warren and Louis Brandeis in a famous *Harvard Law Review* article from 1890. Back then, the two lawyers railed against "recent inventions and business methods" such as "instantaneous photographs" and "numerous mechanical devices" that "invaded the sacred precincts of private and domestic life." What would they think of smartphones and their systematic abuse of our attention and invasion of privacy? I saw myself as a disconnection crusader, a Don Quixote for the digital era, toiling against the tyranny of always-on mobile devices. (Never mind that Don Quixote was delusional.)

My mission was as futile as fighting windmills. Cellphones hardly existed two decades ago. By 2019, eight in ten American adults owned a smartphone; in my own demographic of Americans aged thirty to forty-nine, 92 percent owned smartphones. By 2020, 5.2 billion people worldwide owned a cellphone. Whenever I walked into a public restroom, a guy at the neighboring stall held a smartphone in his free hand. A colleague so vigorously swiped and typed on her iPhone that she injured her wrist and came into the office wearing a brace. My mother, a public school teacher, was encouraged to tweet from the classroom. My father, a minister, contended with congregants answering their phones during church services. Jenna carried two smartphones, one personal and one provided by her employer so she could be reached any time of any day. Seven decades after Congress set the workweek at forty hours through the Fair Labor Standards Act, it seemed time to establish new rules to prevent our jobs from pervading our lives via smartphones. "You can't miss nobody in 2017," the comedian Chris Rock said during a stand-up routine that year. "Not really. You can say it, but you don't really miss the motherfucker, because you're with them all the time. They're in your fuckin' pocket."

My refusal to swim with the digital current made me an outsider, a fringe character unable to accept the inevitable march of technological progress. Without a smartphone, I couldn't use Uber, Venmo, or WhatsApp. By choice, I also opted out of Instagram and most other social media. When I started a fellowship at the Columbia University Graduate School of Journalism in 2016, I was instructed to join a Facebook group to stay updated on school events. When I told a vice dean that I wasn't on Facebook, she rolled her eyes and asked how anyone expected to be a journalist if not on Facebook.

Even access to basic needs has started to hinge on having social media and a smartphone. Cities have been swapping out traditional parking meters in favor of mobile pay-only zones. In 2018, San Francisco began requiring a download code to use some public bathrooms. The same year, credit reporting companies began using cellphone plan payments to determine credit scores. I was once refused take-out service because I couldn't call in my order, even though I was standing at the restaurant's door. I felt like a recluse, a modern hermit in plain sight.

If I were to try to psychoanalyze myself, I might say I was reacting in part to growing up in a conservative Christian home and feeling pressured to be "always on" for others. I had to attend my father's church every Sunday and sit in the front pew beside my mother, an intensely upbeat woman who wanted her children to set an example. Rebelling against the cellphone was, perhaps, a belated way of cutting the cord with expectations for how to behave. I also remember a father who fell asleep in front of the television most nights. I grew up to resent the screen. In college, after reading Bill

McKibben's *The Age of Missing Information,* I vowed to never own a TV. One summer, I unplugged all the wires from the back of our family television to enforce a cold turkey detox. (It only worked because our parents were traveling for a month.) And what is a smartphone but a demonic iteration of a TV screen?

It's not that I want to return to the nineteenth century. I appreciate the death of corded phones; in my childhood kitchen, the curly cord of our wall-mounted phone was always running across the room like a clothesline, threatening to behead passersby. I know to be wary of rose-tinted nostalgia for a "simpler time." I just don't understand why carrying a smartphone should now be a prerequisite for living. I don't want to give others access to every minute of my life. I don't want all my information to come from screens. I don't see the need to be constantly connected and reachable. I am already online enough, on my computer. I don't want to be a person who looks down at a phone midconversation. And I don't want others to be that way, either. So much of the digital world was designed to make us feel dissatisfied, to mine our thoughts for marketable content that can be sold back in the form of Google ads and Amazon one-click purchases. I don't want to live in that world, even if it means I occasionally get lost driving (or bicycling).

I've come to find that I'm not alone in this crusade. Having lunch at a café a block away from Columbia's leafy campus in early 2017, David Helfand and I were the only patrons without smartphones resting in front of us. "I've occasionally met other people like me, but not many," said Helfand, an astronomy professor and former president of the American Astronomical Society. He had a laid-back demeanor—with white hair and a thick beard, the astrophysicist had literally played the part of Santa Claus—but he expressed zero tolerance for smartphones. His refusal to get a cellphone nearly got him into a legal fight with the federal government.

In 2016, the Social Security Administration announced that senior citizens would henceforth need a cellphone to access their Social Security accounts; entry to the website would require a two-step verification involving a passcode sent to a cellphone. "I write to register my outrage at your new policy," Helfand wrote to the administration.

> Has it not occurred to you that some people in this country cannot afford a text-enabled cellphone? Has it not occurred to you that some people live in areas without cellphone service? And has it not occurred to you that some people, with plenty of money and access such as myself, might actually choose not to partake in the toxic cesspool of social media, and might value the ability to manage their own time, deploying their mental resources on topics with more substance than tweets and the rest of the superficial banality that passes for "conversation" today?

Two ranking leaders of the Senate's Special Committee on Aging called for reconsideration of the policy. The Social Security Administration backtracked. Helfand had prevailed—for the moment.

"I believe that access to me should be at *my* discretion and not at someone else's discretion," Helfand told me. He maintained that life without a cellphone was more efficient. It gave him "freewheeling" time to let his mind wander. It allowed for uninterrupted focus. It created quiet.

That said, Helfand did have a Twitter profile because a student had encouraged him to use it to promote his book, *A Survival Guide to the Misinformation Age*. The student tweeted whatever Helfand approved over email, a setup reminiscent of how some Orthodox Jews, to honor the command to rest on the Sabbath, hire others to do certain labors. Helfand's quasi-religious opposition to social media and cellphones set the stage for tensions. He had walked out on meetings if people were distracted on their smartphones. He had also walked in the rain because he couldn't call his wife for a ride home from the train station. Similar challenges arose when flying. Rather than saying, "I'll call you when my plane lands," Helfand had to say something like, "Meet me at pillar thirty-two at 4:20 P.M."

Helfand may sound like a stubborn crank, irrationally unwilling to make his life (and others') easier, and only the latest in a long line of misguided Luddites. Socrates had opposed the written word because he thought it would undermine memorization. Thoreau had dismissed the telegraph because he thought far-flung places "have nothing important to communicate." There is inevitably pushback to any new technology. But aren't smartphones fundamentally different? Rather than being a tool of the owner, the smartphone controls the user with addictive apps that allow third parties to mine data and sell ads. Amid a wave of social media-undermined elections, smartphone-enabled erosion of in-person conversations, and an infuriating loss of quiet due to always-on devices, what kind of cultural shift could happen if we all started acting a bit more Helfandian?

I once mentioned my decision to live phoneless to the hyperconnected founder of an online news start-up valued at $30 million. (It later sold for multiple times that amount.) He had two smartphones stacked by his side. His business depended on news consumption on mobile devices. Yet he said that if he could live in a world with or without cellphones, he would choose the latter. Then he shrugged, because that was not a real option.

But what if, somewhere, living cellphone-free *was* an option?

What if there *was* a place where people weren't constantly talking on their phones or distractedly scrolling—not during dinner, not during work, not in bed? A place where forest hikes and sunset vistas weren't tainted by a ringtone or the obnoxiousness of a person yapping on a phone? Where getting lost meant *really* getting lost, because a gadget wasn't going to rescue you? A place where GPS apps went haywire, and people floated above the digital currents that glued the rest of America to a screen?

Those questions led me into Appalachia, over snowy mountain passes and down steep switchbacks, into the rugged backcountry of Daniel Boone and Stonewall Jackson, to the heart of the National Radio Quiet Zone, in search of an alternative to our tech-obsessed, phone-addicted, attention-hijacked, doomscrolling society. After my initial visit to Green Bank with Jenna in early 2017, I returned over the next three years for a series of extended stays that would total about four months on the ground, popping in so frequently that people asked if I'd moved there permanently. I was hoping to discover a better way of life, perhaps one that I would want to adopt. I joined a book club, helped build a house, foraged for ramps, and went target shooting with a seven-year-old. I frequented a small country church where the wall-mounted "Register of Attendance and Offering" was never updated; it always said there were eleven attendees and seventy-nine dollars in tithes, contributing to the feeling of time standing still, of being drawn into a quieter dimension. A technician at the Green Bank Observatory would tell me that the Bible actually referenced radio astronomy in Psalm 19:4: "Their line is gone out through all the earth." To him, this was a nod to the radio frequencies given off by all things, making the observatory's work a spiritual endeavor, and Green Bank a godly place.

Others believed fantastical things happened in the Quiet Zone, that it was a transcendent realm where space-time cleaved open and spirits from another dimension entered our world. That it was on one of the so-called ley lines that demarcated Earth's mystical energy fields. "This area is on a node, an intersection, *Twilight Zone* shit," a ginseng hunter named De Thompson would tell me with a throaty laugh, lips curling over his toothless gums. There was a beguiling confluence of supernatural and provincial, of conspiracy and reality, swirling in the thick fog that enveloped the radio telescopes at dusk.

At the start of my journey, however, it did not occur to me that a community bathed in quiet could be anything but idyllic.

CHAPTER TWO

"One of the Science Capitals of the World"

THE FLURRY OF A BANJO'S TWANG kept time with the moan of a fiddle and the strum of several guitars, all of them speeding along to an old-time standard. I sat in the music circle holding a guitar, but my hands were frozen. I'd naively thought I could play along with a bluegrass band called Juanita Fireball and the Continental Drifters, whose fiddler was the principal scientist at the nearby astronomy observatory. Turned out, he was also a state champion banjo player. The group plucked and strummed with such speed and intuition that I gave up after a few songs, red in the face. Green Bank was no place for a mediocre musician.

I slinked off the porch and grabbed a beer from a cooler, mingling with several dozen people at the backyard party. I overheard someone say, "If my cellphone was working, I'd look that up." But she couldn't, because of the radio quiet. As the sun set, the brightest thing in the sky became the blinking red light atop the mammoth Robert C. Byrd Green Bank Telescope looming in the distance, as normal to everyone there as the moon rising overhead.

It was late May 2017, and I'd just arrived for a monthlong stay at the observatory, renting an apartment typically reserved for visiting scientists. My room key came in a white envelope with a piece of paper titled "We have met the Enemy, and He is Us," followed by a list of Quiet Zone regulations. The message was clear: I was the enemy of quiet, tolerable only if my electronic inclinations were kept in check.

"Prohibited devices include cordless telephones and wireless networking devices," the regulations stated. "A good rule of thumb is, if it uses RF [radio frequency] energy to communicate, it almost certainly exceeds the ITU-R RA.769 limit, and we ask that you please don't use it here." That meant my iPod was to be kept on airplane mode and my laptop turned off when not in use. Other types of electronics with strong electromagnetic emissions, such as air-conditioning units and microwave ovens, were confined inside shielded enclosures. All electric cables were buried underground. A plan to install automatic toilet flushers had been nixed after the observatory realized the sensors emitted radio noise. Even the autonomous driving capabilities of an employee's Tesla were disabled.

The rules were strict, but I had to wonder: Who was enforcing them? I had expected a military-style debriefing and thorough inspection of my belongings. But all I got was that letter. After reading through the regulations, I'd changed into running shorts and jogged down a wooded trail that wound around the 2,700-acre federal property and looped by the Green Bank Telescope. Beyond it, past several fields, I could see a white farmhouse. I was later told that it was the residence of Bob and Elaine Sheets, who lived at the very heart of the National Radio Quiet Zone—though they were far from the "quietest" household in town. In fact, they were hosting the biggest party in Green Bank—and likely the biggest thing for fifty miles in any direction—that Friday. Which was how I ended up in their backyard with a beer in hand.

Green Bank didn't have a mayor, but if it did, Bob would likely have been elected. Tall and trim, with a long face that easily broke into a wide grin, he had a youthful energy that belied his sixty-eight years. He wore two rings: a wedding band and a large gold ring from being on the Fairmont State University basketball team that made it all the way to the NAIA championship game in 1968. (Fairmont lost by three points.) After college, Bob returned to Green Bank to teach high school English and coach basketball for thirty-five years, the kind of person whom students called "Coach" both in and outside the classroom. Gregarious and personable, he served on various community boards and regularly hosted events at his farm—including the following morning, when more than one hundred people would be coming to tour one of Green Bank's original settlements. Bob was distantly related to William Warwick, an early settler whose property in 1774 became the site of a colonial militia fort. The fort's exact site was unknown until Bob—with the help of two professional archaeologists and aided by oral tradition—began digging it up in his hayfield. With the use of a metal detector and magnetometer, which Bob got approval to use from the observatory (as the devices threatened to pollute the radio quiet), he'd unearthed a trove of artifacts: a soldier's cuff links, a glass watch fob bearing the image of King George III, and a fragment from a bottle of Turlington's Balsam of Life, a potion for treating "inward weakness."

The area's history also decorated the inside of Bob's home. In his living room was a framed box of several dozen arrowheads dating back as far as ten thousand years, some likely left behind by the Shawnee who once hunted on the Sheetses' ninety-seven-acre property. A confusing point about the area's indigenous history was the county's name, Pocahontas, as there was no record of the Native American woman having ever visited. According to Bob, Pocahontas had descendants who lobbied state legislators in Richmond to name the county after her in 1821, back when West Virginia was still part of Virginia. In any case, Bob said it was because of Pocahontas that the high school's sports teams were known as the Warriors. At the school entrance, a life-size, bare-chested, headdress-wearing "warrior" was poised with a bow and arrow, ready to shoot enemies at the door. Outsiders beware.

A LIFELONG GREEN BANKER. Bob was eight when the National Radio Astronomy Observatory (NRAO) broke ground in town in October 1957. "If you think it's quiet now," he told me, "it was *really* quiet then."

A new and exciting field, radio astronomy had been discovered only a few decades earlier. Karl Jansky, a scientist at Bell Telephone Laboratories in New Jersey, realized in 1933 while testing an early transatlantic radio communication system that he was detecting radio signals from space. "New Radio Waves Traced to Centre of the Milky Way," the front page of the *New York Times* heralded on May 5, 1933. Hearing of the discovery, an electrical engineer named Grote Reber built the world's first radio telescope in 1937. In his mom's backyard near Chicago, in what might be the quintessential DIY science project, he assembled a thirty-one-foot-wide parabolic dish of sheet metal that focused incoming radio waves onto a single point called a receiver, which amplified the incredibly faint signals into recordable measurements. As Reber aimed the dish at various points in the sky, the receiver picked up different strengths of radio waves, allowing him to create the very first "map" of the radio universe, similar to how an optical astronomer creates a map of the visible sky. But while an optical astronomer is limited to viewing visible electromagnetic radiation (or light) in the frequencies between 400 terahertz and 900 terahertz, the radio astronomer observes lower frequencies between about 1 megahertz and 1 terahertz. (Microwaves are considered a subset of radio waves. Higher frequency waves in space, such as infrared rays, x-rays, and gamma rays, are blocked by Earth's atmosphere and can only be measured from satellites.) After two years of work, Reber published his findings in *The Astrophysical Journal*. The science of radio astronomy was born.

The field received a major investment in 1956, when President Dwight Eisenhower requested $7 million from Congress for the creation of "the nation's first major radio astronomy center" under the umbrella of the recently formed National Science Foundation, which had already embarked on a multiyear search for the quietest suitable place to host such an observatory. According to a 1954 meeting of the NSF's Advisory Panel on Radio Astronomy, the site needed to be largely free of radio noise, able to host a dozen major telescopes and campus facilities, in a sparsely populated area surrounded by mountains, and at least fifty miles from the nearest city. The site would preferably be in a northern latitude, yet far enough south to allow for observation of the center of the Milky Way. Snow and ice were to be minimal. The site also needed to be within three hundred miles of Washington, D.C., and "easy to reach by plane, rail, or automobile." Of thirty short-listed locations, Green Bank came out on top. It was hemmed in by Cheat Mountain and Spruce Knob, the two highest points in the state, each peak nearly five thousand feet high. It was only two hundred miles from Capitol Hill. The surrounding county's once booming logging industry had gone bust, leaving behind ghost towns.

"Green Bank Assured of Great Astronomy Center," the *Pocahontas Times* declared in July 1956. "And How Truly Thankful Are We All, Too!"

Not quite *everyone* was thankful. Some farmers felt pushed off their land as the Army Corps of Engineers, with $550,000 in acquisition funds, began amassing 2,700 acres stretching across the hamlets of Green Bank and Arbovale. The *Charleston Gazette* photographed three Green Bank women by their hundred-year-old homesteads, each angered at being forced to sell the family inheritance. "What will we do?" asked one. "We've worked hard here. It's our home. Where will we go?" A government lawyer initiated condemnation proceedings against at least four families who refused to sell. Resentment would simmer for decades. Kenneth Kellermann, a longtime astronomer at the observatory, said local people "gave the impression that the troops marched in, tore babies from their mothers' breasts, pulled people out of the beds that they had been born in, and forced them out of the house."

Nevertheless, according to the *Gazette*, there was excitement for a facility where "the smallest telescope will be higher than the Daniel Boone Hotel"—a ten-story luxury accommodation in Charleston. "It's the biggest thing that's happened here since Bruce Bosley made All-American," a local told the newspaper, referencing a Durbin boy who played football for West Virginia University and the San Francisco 49ers. The *Inter-Mountain* newspaper of Elkins declared that Green Bank was transforming "from a little farm village of 100 population to one of the science capitals of the world."

State authorities were thrilled with the influx of federal dollars. The West Virginia Legislature showed its support by holding a special session in August 1956 to enact the Radio Astronomy Zoning Act, which created legal protections within ten miles of "any radio astronomy facility in the state of West Virginia." It remains the only state law of its kind in the United States. To further protect the area, the Federal Communications Commission in 1958 created a surrounding thirteen-thousand-square-mile National Radio Quiet Zone, the first of its kind worldwide.

One of the observatory's early hires was Bob Sheets's mom. For more than three decades, Beatrice Blackhurst Sheets was the executive secretary and welcome wagon for astronomers from around the world who traveled into the Appalachian Mountains to unlock the mysteries of the universe using some of the world's premier radio telescopes. One day while visiting his mom at work, a preteen Bob passed an office where an astronomer sat staring into space. When Bob passed the door again hours later, he noticed the man hadn't budged. Bob asked his mother, "What does Dr. Drake do?" She explained that Frank Drake was a scientist. Bob replied, "I walked past his office twice today and he hadn't moved. It didn't look like he was doing anything." The Harvard graduate was paid to think, she said. "He gets paid to think?!" Bob responded. "Wow, that's a great job!" He would have been even more shocked to know that Drake may well have been thinking about how to communicate with aliens.

As the first telescopes rose from Green Bank's hayfields, Drake turned them toward the cosmos in search of E.T. "In a limited way," according to an October 1959 press statement from the observatory, "the National Radio Astronomy Observatory has commenced a project, called 'Ozma,' whose purpose is to detect radio emissions created by intelligent beings on other planets." (Ozma was the name of a princess in L. Frank Baum's Oz book series.) Drake dubbed this new branch of science SETI, or the Search for Extraterrestrial Intelligence. Pretty quickly, in 1960, he picked up what seemed like an alien signal. It turned out to be human interference, perhaps aircraft radar. Otherwise, he detected only static.

Undeterred, the following summer Drake hosted a "quiet meeting of scientists qualified to discuss the scientific aspects of establishing radio contact with other planetary systems," as it was phrased in an internal memo of the National Academy of Sciences' National Research Council. Attendees included the cosmologist Carl Sagan, who later wrote the novel *Contact* about an astronomer working on a project called SETI at the National Radio Astronomy Observatory; the biochemist Melvin Calvin, who won the 1961 Nobel Prize in Chemistry while in Green Bank for having discovered the stage of photosynthesis when plants convert carbon dioxide into glucose; and John Lilly, a neuroscientist famous for attempting to establish human-dolphin communication—a couple years later, Lilly developed an experiment where a young woman lived in a flooded apartment with a dolphin named Peter in an effort to teach him English. The group formed a semisecret society called the Order of the Dolphin, on the idea that if humans could learn to speak with dolphins then we might communicate with aliens, too. Drake presented a now famous equation that calculates the likelihood for intelligent alien life. Amid philosophical debate about our species' tendency toward self-destruction, Drake estimated that ten thousand technologically advanced civilizations existed in our galaxy alone. The Milky Way is one of two trillion galaxies in the universe.

There was some irony to Green Bank's history as the founding place of the still-ongoing search for extraterrestrial intelligence. Since then, while E.T. has continued to elude humanity, Green Bank itself has become more and more alien to the rest of the world. Protected by the Quiet Zone, a place with restrictions on technology and without 24/7 connectivity in the modern age is today so strange, so foreign, so *bizarre* as to seem exotic. By 2017, the observatory was hosting around thirty media visitors a year and taking more than one hundred press inquiries annually, with a regular stream of articles being published about the Quietest Town in America. "Enter the Quiet Zone: Where Cell Service, Wi-Fi Are Banned," National Public Radio reported in 2013. "Life in the Quiet Zone: West Virginia Town Avoids Electronics for Science," *National Geographic* wrote in 2014. "America's Quietest Town: Where Cell Phones Are Banned," CNN said in 2015. Busy days could see three film crews crowded on the Green Bank Telescope at the same time, competing for footage of that most endangered of things: quiet.

SETTLING INTO GREEN BANK, my mornings often started with a jog around the observatory's wooded property, where I was encountering greater diversity of animal life than I'd ever seen outside a zoo. I nearly stepped on a rattlesnake, spotting its coiled body just in time to leap over it. I passed a fawn that was curled in a ball, pretending to be invisible. A coyote popped out of the woods and trotted ahead of me for several hundred feet. I rounded a bend and saw two cubs scurrying up a tree, which sent me running the other way for fear I'd anger the mama bear.

I quickly felt like I knew half the town. Because there wasn't much to do and there weren't many people, I was seeing the same people at the same events: at Bob's house, at a community breakfast of biscuits and gravy, at a square dance, and at Trent's General Store, where I got gas and groceries. The fuel pumps had no credit card reader, which meant everybody paid inside, where the wood floors were worn to a sheen and customers traded the latest gossip about who'd fallen off a four-wheeler or killed a bear.

The observatory straddled the hamlets of Green Bank and Arbovale, each with its own general store: Henry's and Trent's, respectively. I quickly became a Trent's regular because of its proximity to my apartment, a half mile away. Two old grocery store conveyor belts—purchased secondhand from a Kroger supermarket in the 1970s—ran alongside two cash registers, each covered in sticky notes, some with memos to be passed along to shoppers. Because people couldn't call each other on cellphones, they used Trent's to pass messages. It was a community hub. Longtime customers paid on account. Fresh meat was sold next to a refrigerated case for fruits and vegetables restocked every Thursday. Aisles were packed with nonperishables: cans of beans, jars of pickled pigs' feet, bags of Atomic FireBall candies. You could buy a hunting or fishing license, plus ammunition or bait. Hanging from the walls were a selection of turkey callers and an air rifle packaged in a box with a picture of a squirrel. A cabinet held livestock medicine. The store also offered dry-cleaning services; clothes wrapped in plastic hung at the front for pickup. Cereal cost about five dollars a box, twice as much as at a Dollar General two miles away, but the cereal at Trent's came with local charm.

I'd drop by for the newspaper and a conversation with Betty Mullenax, who'd been ringing up customers for a half century, always wearing a blue smock, with her short, curly hair always neatly combed. Her husband, Ebbie, sliced up meats in the back while chewing tobacco and swallowing the juice. They'd married in 1953 and taken a stake in Trent's General Store in 1965 from namesake Omar Trent, purchasing it outright in 1973 when Trent retired. In 1992, they passed ownership to their daughter Debbie and her husband, Bobby Ervine, who expanded with another gas station and store a couple miles up the road, creatively called Trent's 2 and Trent's 3.

"Why is it still called Trent's?" I once asked Betty.

"It's too much trouble to change the name," she said.

Trent's was closed on Sundays, Thanksgiving, and Christmas, but otherwise Betty and Ebbie never took a day off. Ervine managed the business, often wearing a Budweiser cap and with a lit cigarette in hand. "I told the FDA, 'When you start paying my property tax then you can tell me not to smoke in my store,'" said Ervine, who was thin and wiry with an unkempt beard. When I asked what the most exotic item sold in Trent's was, he responded, "Other than moonshine?" and led me to an unmarked cooler containing several two-quart Juicy Juice jugs. "You like berry or apple?" he asked. I opted for berry. Back at my apartment, I poured myself a tall glass of authentic Appalachian moonshine on ice. I nearly gagged, it was so sickly sweet. Ervine also offered to cook me rattlesnake, which I declined.

The store's semiofficial slogan was "If Trent's doesn't have it, you don't need it." It sounded like a Quiet Zone proverb, translating as "If we don't have what you *think* you need, then maybe you should reconsider your needs." It reflected how residents took pride in making do with less, including cell service. "Never had it, never missed it," numerous people would tell me, as if they'd all rehearsed the line. "This is probably the sixth time I've been interviewed," said a teenager named Mathias Solliday. "I've been on the *Today* show. I was interviewed by a couple of Germans who were at Trent's. They all ask the same questions. I tell them all the same thing. I love it here. I don't miss cellphone service."

Aside from no cell service, living in Green Bank meant no malls, no bowling alleys, no movie theaters, no delivery pizza—not because of Quiet Zone restrictions, but because the county was too sparsely populated to support them. The lack of such conveniences was why Michael Holstine, the business manager at the observatory, was initially concerned about moving to Pocahontas in 1992 from the city of Fairmont with his wife and three daughters. He thought, "Shucks, if I bring them here there's a lot they won't be able to do. You can't just go to the movies on Friday night willy-nilly." In time, Holstine's daughters discovered other things: hunting, fishing, skiing, horseback riding, nights so dark they could see the U.S. Space Shuttle docking with the International Space Station against a backdrop of constellations.

"They're glad they were raised here because they feel like they know more about life," said Holstine, a barrel-chested man with a graying beard. "They know how to use their electronic devices. But the difference is that when they come home, it doesn't bother them to put it away. Whereas I think a lot of kids who have grown up with it, they can't do without it—it's like cutting off their hand."

Person after person began telling me stories like this, and I believed them, though I also felt they were trying to sell me a tidy image of rural life. David Rittenhouse, a longtime Brethren minister who had started several churches in the area, called Pocahontas "one of the best places in the world to raise kids." Sure, his son Julian was nearly bitten by a rattlesnake at age twelve while foraging for ginseng. Yeah, a neighbor's dog had killed all but one of the dozen offspring of his beloved Spanish fighting cock, which his wife had smuggled as an embryo into the United States from Ecuador, where they'd worked as missionaries. And to be sure, another neighbor's dogs had repeatedly broken into Rittenhouse's fields and killed dozens of his sheep and calves. But it was all better than being crowded into a city, he argued.

"The loneliest people in the world live in cities, if you ever think about it," Rittenhouse told me. "In the city, a neighbor is a threat and competes for your space. Here, we're scattered out with only nine people per square mile, so somebody five miles away is an asset. Psychologically and socially, we're closer than the people who are living in the city piled up on each other."

Rittenhouse was an outsider—he and his wife had grown up in Maryland and Virginia, respectively—but he was perhaps the most accepted outsider in Pocahontas, owing to his role as a longtime minister, his son Julian's role as a minister and Green Bank teacher, and his grandson Abe's role as a minister, high school teacher, and sports coach. The Rittenhouses were said to marry everybody and bury everybody in the county. David Rittenhouse himself had performed more than one thousand funerals as well as hundreds of weddings. One couple was married at the observatory. Another wanted to be married in a cave where a dead body was found in 1975.

"The interesting thing about living here is you're free to be a character of any kind," Rittenhouse said. "You're not expected to conform and all be the same."

If there was an old-fashioned intimacy to the community, there was also a feeling of insularity. Everyone seemed to know everything about everybody. It was impossible to go unnoticed. Before long, I caught wind of a rumor that a stranger was running around town in a bikini. I hadn't realized how scandalous it was for a shirtless man in V-notch shorts to jog in the Appalachian countryside.

EVERY EVENING, before Trent's closed at 7:00 P.M., Ebbie Mullenax shuffled up to his wife's register and put the cash in an old cloth sack that he stowed in a safe overnight. Their son-in-law Bobby Ervine usually locked the front door and followed Ebbie and Betty out the back as they walked over a well-worn path through the grass to their house next door. They planned to work as long as they humanly could—or as long as the store stayed open. That depended on the observatory.

"I'd hate to see it go out," Ervine said of the science facility. "Anything happens over there, we'd have to close up. Because that's all there is."

Over the decades, scientists from as far away as Russia and Japan had popped into Trent's for supplies. So had at least one millionaire: Jay Rockefeller, the six-foot-seven former governor of West Virginia and a two-decade-long U.S. senator. One of the most notable politicians from one of America's wealthiest families, Rockefeller was the great-grandson of the oil tycoon John D. Rockefeller Sr. and the nephew to former vice president Nelson Rockefeller. For years, he and his family had swung by Trent's for groceries and fishing supplies before driving on to their nearby estate.

Rockefeller retired from politics in 2015 and lived full-time in Washington, D.C., but he still had an account at Trent's, his name handwritten on an index card inside a wooden box beneath Betty's register. I asked why a millionaire who could have lived anywhere chose Green Bank.

"Because it's quiet," Betty said. *Obviously.*

Too quiet, for some. Within a decade of the observatory's establishment, Green Bank had a handful of world-class telescopes of unprecedented size, but life in Appalachia proved challenging for the newcomers. Health care was an issue; on at least two occasions in the early '60s, wives of observatory personnel delivered babies en route to the nearest full-scale hospital, fifty miles away. Education was a concern. In an August 1962 letter, NRAO director David Heeschen called local schools "our biggest problem, and it offers a real threat to the permanence of the NRAO at Green Bank." Scientists' spouses had trouble finding employment. In response, in 1965, the NRAO relocated its administrative headquarters to Charlottesville, 120 miles away. Locals felt snubbed by the exodus. An editorial in the *Pocahontas Times* poked at the scientists' "desire for urban so-called advantages and luxuries."

The telescopes needed to remain in a quiet place, so Green Bank retained a number of scientists and support staff, including Kellermann, the astronomer, who arrived in 1965 in a red Chevy Corvair convertible, top down. He'd grown up near Coney Island, Brooklyn, and earned his Ph.D. from the California Institute of Technology outside of Los Angeles. Now his only entertainment options were a television with three channels and a drive-in theater that showed old westerns. "The audio was terrible," Kellermann said. A tiny speaker would sit in the window of the car, connected by a wire to the sound system. "If you weren't careful and you drove off, you'd break the wire."

By the time I arrived, the theater had shut down and Rockefeller's estate was up for sale. The community seemed to have grown that much quieter.

CHAPTER THREE

"Lots of Bad News"

THE WORLD FALLS STILL when you're suspended nearly five hundred feet in the air. I was standing atop the tallest man-made structure in West Virginia, having taken two elevators and traversed a series of catwalks to arrive at a small platform above the 2.3-acre dish of the Robert C. Byrd Green Bank Telescope.

"What's going on up here?" Bob Anderson, the head of telescope operations, asked a handful of engineers crowded on the airy platform.

"Lots of bad news," one man responded. "Looks like the feed blower motor's broken."

The engineers stood around a motor that powered a hot air blower, which played a supporting role in the telescope's efforts to "hear" the faint waves of electromagnetic energy emitted from stars, pulsars, and cosmic events light-years away. When those radio waves hit the big dish, they bounced up nearly two hundred feet to a subreflector dish above our heads; the subreflector concentrated the waves down through a window at our feet and onto a receiver that converted the waves into electronic signals. The hot air blower prevented ice from forming over the window. If ice formed, the signal would be blurred. On a telescope of this size, it seemed as if a million things had the potential to act up at any moment.

Stretching out below us was the telescope's clam-shaped dish, 330 feet by 360 feet, made of 2,004 aluminum panels, each painted a special matte white because a glossy coat would reflect so much sunlight as to cook the radio receivers, like a magnifying glass concentrating sunlight on an ant. Because the dish flexed with the changing temperatures, twenty-seven sensors gauged the expansion of the panels and communicated that information to a computer that commanded 2,209 actuator motors mounted underneath the dish to adjust the panels. Anderson called this "dynamic corrections." He was even prouder of the seventeen-million-pound telescope's sheer precision. With the touch of a button, the dish could point at any spot in the sky, slowly tracking it as it moved with Earth's rotation. During high winds, the dish tilted horizontally to reduce drag. After a heavy snow, the dish pitched 90 degrees to release an avalanche known as "the big dump."

As the morning fog burned off, surrounding landmarks came into view. I could slowly make out the long concrete building of Green Bank Elementary-Middle School, and then the observatory's 3,500-foot-long airstrip, rarely used since Jay Rockefeller stopped flying into town. Because of frequent cloud cover, landing a plane here could be a harrowing experience, and I was told of pilots being forced to reroute in thick fog and snowstorms. Six other telescopes emerged from the haze, their dishes pointing toward the sky like giant earlobes. Half the scopes were dormant, too old and antiquated to be worth operating. I finally glimpsed the Sheetses' white farmhouse, a basketball hoop in the driveway.

A loud, clanging noise came from below us in the radio receiving room, where radio waves were measured, filtered, and cryogenically cooled to negative 433 degrees Fahrenheit, which preserved their signal and mitigated outside electrons from mixing in. The signal was converted into electricity, then into light, and then transmitted through underground fiber-optic cables to the observatory's control room a mile away, where terabytes of new information every day were stored on five-foot-tall stacks of hard drives. Twenty-four hours a day, a telescope operator sat at a desk with more than a dozen computer screens that provided instant feedback on what the observatory was picking up from deep space. The control room was sealed off with copper-lined walls and copper-screened windows, acting similar to how a screen on a microwave window prevents radiation from escaping. A dreadlocked engineer named Galen Watts, whom I'd first met while playing music on the Sheetses' porch, likened working at the observatory to doing sound for a rock concert, as he had done for bands including the Grateful Dead and Kansas. Just as a sound engineer amplified a rock band's performance into an intimate experience for thousands of people, so did a telescope engineer turn the faint radio waves of space into comprehensible data for astronomers.

Anderson's team would soon have the feed blower fixed. But other problems loomed. To learn more, Anderson recommended that I talk to Jay Lockman, the observatory's principal scientist. I already knew Lockman—he was that damned fast fiddler and banjo player.

ON LOCKMAN'S OFFICE DOOR was a photo of himself with the filmmaker Werner Herzog, who'd visited a few years earlier for the documentary *Lo and Behold: Reveries of the Connected World*, which portrayed Green Bank as the antithesis of a digitized society. Herzog reportedly doesn't use a cellphone and has described life in Green Bank as "America at its best." He included a scene in his movie of

Lockman playing music with his band. On the scientist's desk was an inscribed copy of the book *Herzog on Herzog:* "To Jay," Herzog had written, "let's look together to the deepest abysses of the universe!" The movie mementos in Lockman's office continued, curiously, with a poster of Fred Astaire and Ginger Rogers tap-dancing in the 1935 film *Top Hat.*

"Every time I watch Fred Astaire, it makes me a better scientist," Lockman explained. "He was such a perfectionist. He did things till he got them right . . . It's that kind of devotion to the craft, the intense effort, that I admire. It comes off as being totally flawless, totally at ease. There's no awkwardness."

Lockman's work required a perfectionism that would impress even Fred Astaire. In the 1990s, he oversaw construction of the Green Bank Telescope, which I had climbed atop. It was the largest movable object on land and one of the most sensitive large-scale scientific instruments ever built. It was his great tap dance. On his computer, Lockman brought up an image of the Orion Molecular Cloud based on data collected by the telescope. A flame-orange ribbon, circled by a ghostly halo, bisected a patchwork of stars that formed the constellation Orion, 1,344 light-years from Earth. The orange was ammonia—the same kind used in household Windex—and it gave off a unique radio wave picked up by the telescope. Everything emits electromagnetic radiation, with each frequency a kind of fingerprint for the astronomer to study. "This shows the GBT at its best," Lockman gushed.

A leading authority in his field, Lockman had first visited Green Bank in 1967 while an undergrad at Drexel University in Philadelphia, where he grew up. While later interning at the National Radio Astronomy Observatory headquarters, he had been one of the first to realize that a Green Bank telescope had detected ionized gas at the galactic center, suggesting a black hole in the Milky Way.

Lockman had thought, "I know something that nobody else does!" Such excitement propelled him to a lifelong career in science, and he was at the moment recording a Great Courses series on radio astronomy. Working in Green Bank was "a nice combination of being at the center of the universe and way out on the fringe all at once," Lockman said.

It was a warm summer day, so he suggested a stroll around the observatory's campus. In khaki shorts, Lockman kept a brisk pace toward the telescopes. He pointed to a clump of asparagus growing by a fence. So much grew in Green Bank that he never bothered going to the store for it. He just walked into his yard or down his dirt road and picked a few pounds. A neighbor with the Sheetses—at least as much as you can be a neighbor when separated by a river and a forest—Lockman lived in a renovated farmhouse filled with regional folk art and lacking television or WiFi. Nor did he have a smartphone, which was more than just a default of living in a town without cell service. He loathed the device, comparing it to cigarettes in its purposeful addictiveness—something he knew about, as a former smoker. It was a comparison I'd often heard. A few months earlier, I'd interviewed a psychotherapist in Manhattan named Robert Reiner who treated people for tech addiction, and he described smartphones as the modern cigarette—a way to relieve social anxiety. Suggestions I'd read for how to break the addiction included reinstating the "away" message on email, leaving the smartphone at home, or buying a purse too small to fit a phone. Reiner recommended setting boundaries. If his kids' grades fell, he took away their smartphones. And he banned devices from the dinner table.

Lockman had a simpler solution: live in the Quiet Zone. When he left his office, he was offline and unreachable. Work calls and emails couldn't follow him, which freed him from what so many people today report as a major source of anxiety and burnout. He believed that time offline boosted his productivity, giving him headspace to focus and think. This wasn't a new idea. The tech writer Nicholas Carr argued in his 2010 book, *The Shallows,* that "quiet spaces" help people to think deeply, make new associations, draw inferences, and foster ideas. In the 2012 book *Quiet,* Susan Cain wrote that "companies are starting to understand the value of silence and solitude" with the creation of "quiet zones" for uninterrupted workflow. Lockman said the quiet made science conferences more productive in Green Bank; attendees weren't distracted by smartphones or tempted to skip meetings in favor of hitting the town. He reminded me of David Helfand, the sage-like astrophysicist in New York City who had encouraged me in my quest to live without a smartphone. As it turned out, Helfand and Lockman had been officemates as doctoral students at UMass Amherst. Their career paths had led them to very different places, yet they espoused a similar aversion to smartphones and embrace of quiet.

Lockman and I walked up to a gate. An old sign read "U.S. Gov't. Private Property. Authorized Diesel Vehicles Only." Gas-powered engines were not permitted any closer to the Green Bank Telescope a mile away, as their motors relied on spark plugs, which emit a burst of electromagnetic radiation. A weathered sign had the image of a spark plug with a line across it, like a "No Smoking" sign for radio noise. During guided tours of the observatory, tourists were at this point ordered to completely power down all electronics.

We turned into the adjacent Arbovale Cemetery—an ominous place for a stroll, given how the observatory was potentially facing its own end of days. Green Bank had lost key political support in 2010, when West Virginia senator Robert C. Byrd died in office. Considered the longtime "king of pork," Byrd had personally brought more than $4 billion to his state through his leadership of the Senate Appropriations Committee and seniority as the longest-serving senator in U.S. history. More than forty major facilities in the state, from the Robert C. Byrd Green Bank Telescope to the Robert C. Byrd Highway, bore his name. For decades, his staff had called the observatory every year to offer financial support. "He'd say, 'We're getting ready to start our budget appropriations meetings, and do you need money for anything?'" recalled Michael Holstine, the business manager.

No more. Soon after Byrd's death, as part of a decadal review process, a committee of the National Science Foundation announced the observatory may not be worth its annual $14 million upkeep. With an $8 billion budget, the NSF was tasked with keeping America at the cutting edge of everything from geology to zoology, and its annual outlay for astronomical sciences was about $250 million. Other new astronomy facilities were coming online, taking up a greater portion of the budget, and something had to give. Green Bank apparently no longer fit the bill, despite its research having contributed to numerous scientific discoveries as well as advancements in microwave communications systems, data recording technology, image restoration techniques, remote sensing, navigation, and geodesy. In 2016, the NSF announced it was considering mothballing the facility.

The same year, the NSF split Green Bank from the National Radio Astronomy Observatory and made it an independent unit, which gave Green Bank more flexibility over its operations. The observatory shifted away from Open Skies, a program where anyone in the country could apply for free telescope viewing time. The telescopes were increasingly rented out. By 2017, nearly one-third of Green Bank's funding was coming from Russian sources—primarily the billionaire Yuri Milner's initiative to search the universe for intelligent life, a project known as Breakthrough Listen. The NSF still owned everything and funded around two-thirds of the observatory's budget, but that financial support threatened to crater. The future appeared bleak for Green Bank.

Walking among the gravestones, Lockman discussed another existential challenge for the observatory, one rooted in the fabric of modern life: radio noise. WiFi and other forms of wireless communication and gadgetry were spreading like wildfire in the area. Lockman said the Green Bank Telescope had recently detected a new molecule in space but was unable to confirm the finding because of interference from satellite television transmissions. (The National Radio Quiet Zone only protected against interference from ground-based transmitters.) To confirm the molecule's discovery, scientists had to use a radio telescope in Australia, where satellite TV operated on a different frequency, fortunately. "That was a wake-up call for radio astronomy," Lockman said. "You can't just hide in a place like Green Bank."

Scientists have for decades lamented the loss of quiet. In 1979, the National Academy of Sciences' Committee on Radio Frequencies warned that "the electromagnetic spectrum is now so heavily used on Earth that much of its potential value for passive scientific research has already been seriously affected." In 1987, Claud Kellett, program manager for the NSF's National Astronomy and Ionosphere Center, told the trade journal *FCC Week,* "The only quiet zone [in Green Bank] is gradually being whittled away. The day will come when there will be no way to reduce all the noise." In the same article, Dr. Robert Riemer, program manager of the Committee on Radio Frequencies of the National Academy of Sciences, urged that "we have to preserve bands of spectrum in much the same way we have to preserve wildlife and wilderness."

Governments have shown some willingness to make concessions to radio astronomy. In the 1960s, scientists successfully lobbied to ban TV stations from broadcasting UHF Channel 37, which would have interfered with a major radio telescope in Illinois as well as encroached on readings of interstellar hydrogen. The most abundant molecule in the universe, hydrogen has provided astronomers key insights into the behavior of the cosmos, and its radio band is still reserved for astronomy research. Scientists also successfully lobbied to prevent microwave manufacturers from operating in the range of 10.6 gigahertz to 10.7 gigahertz. Such would have allowed microwave ovens to brown meat instead of merely heating it to a limp pulp, but it would have polluted the 10.68 gigahertz frequency allocated for radio astronomy. That's why microwaves operate in the 2.4 gigahertz frequency. So does WiFi. Unlike microwaves, however, WiFi routers are always on, always radiating, always polluting the airwaves.

The observatory was working on ways to cancel out interfering signals—essentially noise-canceling headphones for a telescope. But it was still vital to have the quietest environment possible in Green Bank. Lockman compared it to purifying drinking water. "If you want clean water, you start with the cleanest water you can get, and then you take out a little bit of something," he said. "You don't start with something terrible."

In other words, you can't turn sewage into Evian. And you can't operate a radio telescope in a noisy environment.

GREEN BANK'S FIGHT for quiet rested for a quarter century on the shoulders of Wesley Sizemore, a man so passionate about his job that he was once reprimanded for overstepping his jurisdiction in trying to protect the Quiet Zone. A native West Virginian with a thick beard, ponytail, and glasses, Sizemore had been known as a kind of quiet enforcer, patrolling Green Bank and knocking on doors to tell people to unplug their microwaves or turn off their WiFi routers. But when he retired in 2011, his position was eliminated. Because of budget constraints, his responsibilities were spread among several other employees, none of whom was tasked with policing the quiet full time.

I tracked Sizemore down at his home about eight miles south of the observatory, and as we chatted on his porch he recalled stories from his time as the Quiet Zone's top cop. Whenever he'd found an illegal source of radio frequency interference (RFI), he had gone up to the suspect's house, knocked, introduced himself "very politely," and asked the offender to cease and desist whatever was causing the trouble.

Sizemore once detected noise coming from the Sheetses' house, pinpointing the infraction to a malfunctioning electric blanket. The observatory bought them a replacement. Because RFI was often the result of faulty electronics—a short circuit, an electric arc—some in the community saw Sizemore as a free repairman, be it for a damaged electric fence or a buzzing stereo radio. The U.S. government could also be a source of interference. The National Emergency Airborne Command Post (NEACP), a plane that acted as a "White House in the sky," sometimes flew overhead and emitted a signal picked up by the telescopes. Sizemore raised the problem directly with NEACP, which was initially alarmed that a civilian knew the flight path of their secretive military plane. NEACP agreed to reroute when possible or give Sizemore a heads-up when flying over Green Bank; he also started sharing the observatory's viewing schedule with NEACP so they could avoid the Eye of Sauron in Green Bank.

In 2009, Sizemore detected noise from a Dollar General that had opened in Green Bank: the automated front door was using a microwave motion sensor. The observatory recommended the store replace it with a sensor that operated on the infrared wavelength, which was outside of what the astronomers measured. Then Dollar General wanted to utilize wireless inventory scanners, so the observatory had a suggestion: paint the entire building in conductive lead paint, which would help prevent electromagnetic radiation from escaping. It was the only Dollar General with a black facade I've ever seen. The observatory did similar mitigation with other local businesses. When a Green Bank medical clinic complained that its service was limited without WiFi, the observatory sent its staff to wire an internet jack into every room.

Not everyone was receptive to Sizemore. A neighbor who'd had some of his land seized through eminent domain during the observatory's creation once grabbed Sizemore by the collar and told him to stay off his property. A longtime schoolteacher told me he was forced to get rid of wireless speakers he'd received for Christmas, so he'd given them to his daughter, who lived a bit farther from the telescopes—but it apparently wasn't far enough, because she was also ordered to give them up. "We can make you get rid of your stuff, it's the law," Sizemore told her (in her retelling).

Some people defied Sizemore. Linda Beverage, a guidance counselor at the county high school, told me that her Green Bank home was one of the first to have a microwave in the '80s. "The truck used to pull up and do its little thing and tell my mom that she wasn't supposed to have a microwave," Beverage recalled. "But my mother was a stubborn little lady, and they didn't take her microwave." Microwaves were a real concern. At the Parkes radio observatory in Australia, astronomers were bedeviled for nearly two decades by a mysterious radio signal; in 2015, they finally pinpointed it to the staff's own office microwave oven.

Ruth Bland, the principal of Green Bank Elementary-Middle School from 2003 to 2011, once decided to test the sensitivity of

Sizemore's equipment by bringing a WiFi router into her office. Two hours after she plugged it in, Sizemore rolled up to investigate. Bland hid the router, never revealing that she was the culprit. When Sizemore again detected noise coming from the school, Bland invited him inside to speak to students about why they shouldn't have iPods or iPhones in their lockers. Sizemore showed up again when he picked up interference from a classroom's malfunctioning thermostat; the observatory's technical shop repaired it, no charge. Free repairs aside, it was no surprise that people felt they were being monitored. "It's almost like Big Brother is watching," Beverage said.

Sizemore had bigger things in mind than spying on Green Bankers. He believed he was helping astronomers expand human understanding of the cosmos, which included looking for E.T. Nearly six decades after Frank Drake launched the SETI project, the observatory was still on the hunt, and, in his way, Sizemore had wanted to help.

"Do you believe in extraterrestrial life?" he asked me as we sat on his porch.

"The universe is so big," I said, "it seems unlikely that we'd be the only life to emerge in all of it."

"There's *got* to be life," he said. "What convinced me was seeing life develop at the bottom of the ocean without sunlight, or when you're seeing microbes at the top of Mount Everest in frozen conditions. Life is ubiquitous. Life will evolve anywhere it possibly can. How long have we been around? That's squat compared to the rest of the universe . . . I think there's a good possibility that one day we'll get that radio signal saying, 'We're out here.' What frequency is E.T. going to call on? Don't you need a place where you can access as many of those frequencies as possible?"

CHAPTER FOUR

"Caretaker of a National Treasure"

"IS THERE A CENTRAL OFFICE for the Quiet Zone?" I asked.

"I *am* the Quiet Zone office," said Paulette Woody. "You're looking at all the employees."

I was in the basement of the Green Bank Observatory's office building, several floors below the telescope control room, feeling a bit like Dorothy Gale peeking behind the curtain at the Great and Powerful Oz. A pepper-haired woman in a plaid shirt and jeans, Woody sat at a modest metal desk.

With the official title of National Radio Quiet Zone administrator, she oversaw the thirteen-thousand-square-mile area out of the most unassuming of accommodations. Her cinder block–walled office had one small window looking up to ground level. The room was decorated with photos of her kids and grandkid as well as plastic green soldiers that summer interns had hidden throughout the building as a prank years earlier. Resting on a cabinet was an award from NASA recognizing how the observatory had helped track the Doppler shift of a probe to Mars.

The retirement of Wesley Sizemore had left Woody as the sole remaining employee focused full time on protecting and maintaining the quiet. Whereas Sizemore had gone out on patrols to monitor the Quiet Zone, Woody acted as a gatekeeper against noisy intrusions. On her desk, a computer had a background image of the Green Bank Telescope surrounded by a massive brick wall—the wall being metaphorical. There was no *literal* wall protecting the Quiet Zone.

"Just because we're in the Quiet Zone, it's not like this shield pops up automatically," Woody said, chuckling at the thought. "It's not a *Star Wars* thing, like, 'We're raising shields around the Quiet Zone to stop radio frequency signals!'"

Bureaucracy was what stopped the noise. Nicknamed the Queen of the Quiet Zone, Woody had since 2005 personally reviewed the system configuration of any type of cellphone, fixed radio transmitter, or other licensed ground-based communication over an area nearly four times the size of Yellowstone National Park. Her office wall was lined with tall filing cabinets full of applications for communications installations, documentation of a decades-long battle to preserve quiet.

Before the advent of the iPhone in 2007, the Quiet Zone administrator typically saw twenty to thirty requests a month for the approval of some kind of communications transmitter. In the decade that followed, the figure quadrupled to an average of 120 requests a month and sometimes to as many as 400, predominantly for cell-related installations. "Cellphone technology has exploded," Woody said. "We went from 2G analog to 4G, soon to be 5G. All those generations of things make more of a demand on the cell companies and radio spectrum usage."

Yet there was still only one Quiet Zone administrator, meaning Woody was perpetually buried in paperwork. If a company wanted to install or modify any fixed, licensed communications equipment in the Quiet Zone, it needed to notify Woody of the installation's location, height, elevation, and power levels so she could run an analysis on whether it might interfere with the telescopes. Her ultimate decision was influenced by the topography of the landscape, because a communications system (such as a cell antenna) on the far side of a mountain was less likely to interfere. For that reason, the proposed location had to be precise. Altering the location for any reason—even if just to the other side of a fence, as had happened—meant Woody had to run the whole analysis again. She might spend anywhere from a half hour to years reviewing paperwork, checking and rechecking sight lines, heights, directivity, and strength.

If Woody did approve a system with specified power restrictions, the observatory's Interference Patrol Group later did a site inspection to check whether everything complied with the application. The patrol group was also on the lookout for illegal installations. Once, a patrol group member was driving to the supermarket in Elkins when he noticed that antennas had blossomed overnight on a cell tower. He photographed the site and showed it to Woody, who informed the guilty party about the Quiet Zone requirements. The company promptly turned off the system and submitted the necessary paperwork for her review.

Woody used to be less concerned about quiet. Prior to coming to Green Bank, she did FCC license coordination for a communications provider in western West Virginia, with the goal of finding the tallest point to spray the strongest legally permissible signal in every direction. That background gave her a personal understanding of the quiet abuses happening all around the NRQZ. The zone's eastern edge ran through the city of Charlottesville, and many communications providers had installed systems on a hill just outside the city to avoid Woody's purview. Tower crews sometimes incorrectly installed antennas *on purpose* so as to provide greater coverage. Woody had anonymous sources throughout the government and communications industry who tipped her off to these noisemakers. She called her informants "my little birdies." She reminded me of the spymaster Varys from *Game of Thrones,* plotting to maintain an edge over enemies.

Woody had a unique job in overseeing one of a dozen radio astronomy quiet zones worldwide, with the others all having been modeled in some way after the National Radio Quiet Zone, which was the very first. The Australian Radio Quiet Zone, for example, was created in 2005 and mandated radio coordination for transmitters up to 260 kilometers from the Murchison Radio-Astronomy Observatory. China also had a quiet zone, established in 2013 and later expanded to cover a seventy-five-kilometer radius of the Five-Hundred-Meter Aperture Spherical Radio Telescope (known as FAST). China relocated some nine thousand people who lived within three miles of its telescope, while Australia's quiet zone was in a remote desert with a human population of two. Woody, meanwhile, had an entire town outside her window that she was mandated to keep quiet by order of the West Virginia Radio Astronomy Zoning Act Code 37a and Title 47 of the U.S. government's Code of Federal Regulations, Part 1.924a. Most of the several hundred thousand inhabitants of the greater Quiet Zone were permitted to have cell service and any other kind of electronic device because they were sufficiently distant from the telescopes. But the closer one got to Woody's office, the quieter things got. She saw herself as a steward.

"I am like a park ranger at Yellowstone making sure that nobody throws litter in the hot springs," she said. "I'm a caretaker of a national treasure, it just happens to be something we can't see."

One other federal quiet zone exists in the United States. In Colorado, the eighteen-hundred-acre Table Mountain Field Site and Radio Quiet Zone has been used for conducting radio and electromagnetic experiments since the 1950s, but it is one-tenth the size of the NRQZ and with far fewer protections. In the early 1990s, the National Science Foundation briefly considered trying to create another federal quiet zone around a collection of radio telescopes in New Mexico known as the Karl G. Jansky Very Large Array. But the idea was quickly abandoned.

"The opinion of the people I have consulted on this idea, both at NRAO and at NSF, is that it would not be wise to try this," NRAO director Paul Vanden Bout wrote to Philip Smith, executive officer of the National Academy of Sciences, in a letter dated June 11, 1991. "The FCC has received a number of requests to review the status of the quiet zone we now have in Green Bank, WV, and so far has declined to do so. The Commission would, in all likelihood, vigorously resist the establishment of another." Maintaining Green Bank's Quiet Zone had proven enough of a headache. And that was before WiFi and smartphones.

Pushback against the Quiet Zone regulations sometimes turned into physical confrontations. Soon after Woody moved to Green Bank in 2005, a stranger at church pointed a finger in her face and said, "You're the reason why I can't talk to my daughter on her cellphone." Woody responded, "I just started a week ago and don't know what you're talking about."

"How often do you have to contact the FCC about enforcement of the Quiet Zone?" I asked.

"I try not to give that secret away," Woody said, which I took to mean that she tried not to annoy the FCC. "Our goal primarily is to work with someone if there's an infraction . . . It's like: 'Let me explain to you the requirements, let me tell you coordination procedures, we'll get it all taken care of and as long as you work with me diligently to get this resolved then we're not going to pursue anything other than getting it taken care of, which is what the FCC would prefer us to do.'"

In her fight to protect the quiet, it turned out that Woody had an even more powerful ally than the FCC: the Department of Defense, which happened to also have a major presence in the Quiet Zone. I'd been hearing rumors about a secretive government facility in the area. Speaking with Woody, I realized that she played a key role in its operations.

BACK IN THE 1950s, when Green Bank was grabbing headlines for its new astronomy observatory, the U.S. military had been quietly building its own radio antenna of unprecedented size in a nearby mountain hollow. Nicknamed the Big Ear, it was to be twice as big as the biggest telescope ever built, with a dish six hundred feet in diameter and rising sixty-six stories high. The forty-million-pound beast would rest on wheels that rotated 360 degrees, allowing it to point in any direction. It was an engineering feat on the scale of the Brooklyn Bridge, carried out in top secrecy because it had "certain military applications of high priority," as the *New York Times* reported in June 1959. Built by the U.S. Navy's Naval Research Laboratory outside the West Virginia town of Sugar Grove—about thirty miles northeast of Green Bank—the Big Ear was intended to monitor Russian communications and missile launches by intercepting radio signals that bounced off the moon. The $79 million telescope ($700 million in today's dollars) was projected to require a staff of more than one thousand people.

The proximity to Green Bank was not coincidental. The National Radio Quiet Zone was established to protect both Green Bank *and* Sugar Grove, with the thirteen-thousand-square-mile rectangular area balanced over *both* towns. In fact, the navy had initially wanted to build its Big Ear in Green Bank, but the astronomers snapped up that property first, according to a December 1956 statement from Congressman Harley O. Staggers announcing the

Sugar Grove project. "Although the two projects have no direct connection," Staggers said, "it is felt they will be beneficial to each other because of the nature of the radio and research work."

After laying the Big Ear's tracks and excavating a twenty-thousand-square-foot, two-story underground operations facility with a five-hundred-foot-long tunnel underneath the telescope site, total project estimates ballooned to $200 million ($1.7 billion today). The weight of the scope had also grown to sixty-four million pounds. Amid cost overruns, uncertainty over the structural integrity of the design, and questions about the project's necessity in light of new satellite technology, U.S. secretary of defense Robert McNamara halted work in 1962. By then, $63 million (more than $500 million in today's dollars) had been dumped into the woods, with little to show for it but a lot of concrete.

It was a warning for the astronomers, who had considered building their own six-hundred-foot telescope in Green Bank. "It made us realize it would be hard to build anything that big," said Ken Kellermann.

Sugar Grove faced abandonment, raising the alarm of Robert C. Byrd, then in his first term as U.S. senator. To salvage the military site, Byrd lobbied McNamara and President John F. Kennedy to have the navy relocate its worldwide communications system to Sugar Grove from Cheltenham, Maryland, where it was running into trouble with growing electromagnetic interference from surrounding suburbs and industry. Byrd told the Department of Defense that Sugar Grove was "one of the finest sites available in the free world for the reception of radio signals." The Department of Defense approved the request.

At a May 1969 ceremony activating the $32.5 million facility in Sugar Grove, Byrd boasted that two one-thousand-foot-diameter fence-like antenna arrays, known as Wullenwebers, would serve as "ears" for naval radio communications, receiving messages from ships at sea and naval installations in the Atlantic and Caribbean. Over the following decades, Sugar Grove built a collection of radio dishes and Byrd obtained tens of millions of dollars to expand the Sugar Grove complex. One major investment project in 1984, called "Timberline II," consisted of a $7.4 million modernization of the underground operations building so it could house $75 million in "new computer and associated research equipment being purchased from Lockheed Missiles and Space Corporation of California and expected to be operational [in 1989]," according to a memo from Byrd's office that I found in his official archives at Shepherd University. "The military construction project consists of 11,770 square feet of new space and 50,100 square feet of building alterations for a total of 61,870 square feet." The underground facility was now bigger than a football field, including end zones.

Amid the multimillion-dollar investments, secretive activity began taking place. Since the mid-1970s, top officials from the National Security Agency (NSA) had been frequenting Sugar Grove with a new objective: to use the radio antennas to eavesdrop on all communications to and from the United States, according to *The Puzzle Palace,* James Bamford's exposé of the NSA. It so happened that sixty miles northwest of Sugar Grove, in the West Virginia town of Etam, the Communications Satellite Corporation had three huge receiving dishes processing "more than half of the commercial, international satellite communications entering and leaving the United States each day," according to Bamford. Using the antennas at Sugar Grove—codenamed Timberline—the NSA intercepted messages from Etam. In a 2005 article for the *New York Times,* Bamford called Sugar Grove "the country's largest eavesdropping bug," with dishes that "silently sweep in millions of private telephone calls and e-mail messages an hour." The NSA also utilized Sugar Grove as a secure location for vital backup records to originals stored at the NSA headquarters in Fort Meade, Maryland.

The extent of the spy work was further revealed in 2013 by the NSA contractor Edward Snowden. Sugar Grove's telescopes collected "1.8 million contact events, 500,000 mobile events, and 150,000 SMS events per day," as reported in the Intercept. It was a surveillance system enabled by the National Radio Quiet Zone and overseen by its civilian administrator, who operated out of a humble basement office in Green Bank.

ON WOODY'S OFFICE WALL, a topographical map showed the unique physical landscape around Green Bank and Sugar Grove. Both towns were surrounded by a noise-canceling ring of mountains, although Sugar Grove sat in something resembling a deep cereal bowl while Green Bank rested on a turkey platter. The difference was purposeful. Green Bank's telescopes measured radio waves bouncing in from every angle, so they needed longer sight lines. Sugar Grove's telescopes primarily intercepted overhead satellite communications.

"I can't comment on what Sugar Grove does," Woody repeatedly told me—which was, in itself, a way of commenting on the secretiveness of Sugar Grove's activities.

I noticed a hat on Woody's shelf bearing the initials NAVIOCOM, which stood for Navy Information Operations Command. It came from Sugar Grove. Next to it was a military "challenge coin"—typically only given to members of the military or during official military visits—bearing an image of a 150-foot-wide dish at Sugar Grove. It was a gift from Sugar Grove, underscoring how integral Woody was to that facility's operations. Since 1984, the Quiet Zone administrator in Green Bank had commented on Sugar Grove's behalf to the FCC and National Telecommunications and Information Administration for all unclassified assignments, both federal and civil. (Before 1984, only Sugar Grove could comment on federal applications in the Quiet Zone.) Sugar Grove still ran its own propagation analyses based on data forwarded from Woody, as well as processed all classified radio frequency assignments that required security clearances. Sugar Grove provided Green Bank with equipment and other technical resources to assist processing of all other applications.

Publicly, the Department of Defense rarely flexed its muscle with regard to the Quiet Zone, but its presence occasionally became visible. In the 1990s, when Jay Lockman was the observatory's site director, he opposed a proposal from municipal officials about forty-five miles away in Augusta County, Virginia, who wanted to install emergency communications equipment that would have interfered with the telescopes. Augusta officials said their fire department was unable to respond to 911 calls and law enforcement couldn't call for backup from radio "dead zones." The dispute ramped up, and the Augusta County Board of Supervisors asked its congressman to draft legislation that would reduce the scope of the Quiet Zone. Lockman brought the issue to Senator Byrd, who relayed it to the Department of Defense. The Office of the Assistant Secretary of Defense responded by letter that "potential interference could develop and adversely affect the relay of sensitive intelligence information from our Sugar Grove facility if the FCC suggestions and existing National Radio Quiet Zone procedures are not followed. The Department does not look favorably on potential congressional relaxation of the procedures for the National Radio Quiet Zone." Only then did Augusta County back down and agree to comply with the Quiet Zone restrictions. Sugar Grove was like the bigger, stronger brother protecting its kid sibling in Green Bank.

A new picture was coming into focus. The astronomy observatory was something of a cover for the less publicized reason for a Quiet Zone: spy work.

"Those guys don't like to fly high on the radar, right?" Anthony Beasley, director of the National Radio Astronomy Observatory, later told me of the NSA. "Let's say you have a radio quiet zone, and I want to do some stuff in this radio quiet zone and I don't really want anyone to know what I'm doing or observing. So over here [in Green Bank], I have a partner who is observing all kinds of stuff all the time in different frequencies and is trying to keep this whole area quiet. They're the perfect partner for me, because if they're keeping everything quiet, then I get it quiet by default . . . They [Sugar Grove] have a certain mission. They don't necessarily want people to know what frequencies they're looking at or what they're interested in. The way to maintain that kind of anonymity is to have someone else stand in front of it for you."

That is to say, Green Bank stood in front of Sugar Grove.

CHAPTER FIVE

"Spook Stuff"

WEST VIRGINIA HAS A RICH HISTORY in conspiracy and the paranormal. The town of Flatwoods is known for a 1952 UFO landing. Point Pleasant holds an annual festival celebrating the legend of Mothman, a human-size winged creature that inspired a movie starring Richard Gere. There have been sightings of a devilish, farm animal–eating Wampus Cat; a shrieking, wolflike Sheepsquatch; and a tall, headless being with white skin known as the Grafton Monster. And then there's Green Bank's mysterious astronomy observatory.

The proximity of Green Bank to Sugar Grove fueled wild conspiracies about a secret entanglement between the observatory and the military station, with the scientists wrapped up in something more than just science. Some of the rumors seemed to have a glimmer of truth.

An often-repeated speculation was that Green Bank's telescopes stood guard over an underground fortress that linked to Sugar Grove through the region's elaborate cave network. In fact, more than one hundred miles of caves ran under Pocahontas and neighboring Pendleton County, the site of Sugar Grove. And the government *had* surveyed the area's caves for potential use in the 1940s and 1950s, considering possible bunker sites all along Route 220 between the two towns, according to the book *Raven Rock: The Story of the U.S. Government's Secret Plan to Save Itself—While the Rest of Us Die,* by Garrett M. Graff. In 1947, at the start of the Cold War, a *New Yorker* staff writer journeyed into Pendleton County's Trout Cave with the chief of research for the Army Map Service, which was looking for caves that might be suitable as fallout shelters. A half mile in, a member of the party noted, "This might even make a good office for the President." (Trout Cave would remain primarily occupied by bats. In 2008, the entrance was gated off to reduce disturbances.)

Furthering the intrigue, the government did secretly build at least one massive nuclear bunker in the Quiet Zone. In the 1950s, unbeknownst to anyone without the highest security clearances, President Eisenhower proposed that Congress build itself a fallout shelter in case of an attack from the Soviet Union. In 1956, Senate Majority Leader Lyndon Johnson and House Speaker Sam Rayburn authorized construction of a $14 million ($135 million in today's dollars) bunker capable of holding 1,100 people beneath the Greenbrier resort in White Sulphur Springs, about sixty miles south of Green Bank and at the southern edge of the National Radio Quiet Zone.

The project was top secret, but it's hard to keep a 112,544-square-foot bunker secret, especially underneath a luxury resort. A man who helped pour the bunker's concrete later told the *Washington Post,* "You could pretty well look and see the way they was setting it up there that they wasn't building it to keep the rain off of them. I mean a fool would have known."

After the bunker was revealed in the *Post* in 1992, a few Green Bankers began wondering when they might finally learn the truth about their astronomy observatory.

"Of course," said Bob Sheets, the observatory's neighbor, "there's always been a sentiment among people in the community and from around the state that there's something going on over here—that there's something buried under the mountain, that there's subterranean tunnels all under here, spook stuff. And that's because of the bleed-over from Sugar Grove, because they really *do* do spook stuff. And then you throw in the fact that the congressional bunker was built down at the Greenbrier, and you've got fruitful turf for conspiracy theorists."

Locals speculated that Green Bank's telescopes were anything and everything from a front for CIA operations to a cover for missile silos. After all, was it just for the sake of science that the facility operated as a self-contained village with its own water tower, airport runway, residential village, machine shop, and independent plumbing and electrical system backed by diesel generators that could supply power for weeks? "I could tell you what I know, but I'd have to kill you," David Jonese, the county sheriff from 2008 to 2016, said to me with a smirk over dinner in Green Bank. He rattled on about a clandestine facility behind the observatory and a shady nexus with Sugar Grove. He sounded like the police chief from *Stranger Things* trying to figure out what in God's name was actually happening at his town's government facility.

A former mayor said the telescopes likely did a "little spying" on foreign countries. A woman said the observatory could "zap" your home to kill its WiFi. A retired state police officer repeatedly called the observatory to ask about the alien held captive in a secret dungeon on site. The observatory once got a call from a mother asking why her television was flashing a message that read "NRAO," which she presumed was a signal from the telescope. Someone had to break it to her that the acronym stood for "Not Rated—Adults Only," which had appeared because her son was trying to watch porn. People blamed the weather on the observatory, saying scientists could make it rain on command. One resident accused the observatory of sucking radio waves out of the atmosphere, which was the real reason for no cellphone service. Visitors regularly asked, "Are you listening to the voice of God?"

Observatory staff sometimes played along with the more comical conspiracies. Anthony Minter, head of telescope science operations, was once asked by a tourist if any extraterrestrial beings were kept on site. Minter replied, "Did you ever consider that there were fifty other areas before Area 51?" The tourist's jaw hung slack as Minter walked away. Another time, Minter got a call from a man in Tennessee asking if the observatory was emitting radiation or "death rays" toward his state as he'd been "knocked unconscious" several times recently. Minter responded, "I don't think so, since West Virginia isn't scheduled to play Tennessee this year."

A high schooler told me that an observatory elevator went down fifteen floors to a sixty-mile-long tunnel that connected to the Greenbrier's bunker. For evidence of something suspicious, I was pointed toward the trefoil radiation symbol and the words "Fallout Shelter" at the base of the 140-foot telescope, completed in 1965. Inquiring about it, I was told that the basement had eighteen-inch-thick concrete walls to support the telescope's 2,500-ton dish, which rested upon a massive ball bearing 17.5 feet in diameter; it was the world's largest polar-aligned telescope. The walls were also thick enough to qualify the basement as a nuclear bunker. During the height of the Cold War, a couple of handheld Geiger counters were kept inside, along with a stash of food and water.

Over at Sugar Grove, meanwhile, what was the point of that multimillion-dollar underground "operations building"? It was more than half the size of the Greenbrier bunker, with two-foot-thick concrete walls—ostensibly to prevent radio frequency interference, but conveniently suitable for a fallout shelter. Could it really be a nuclear shelter? Part of the rumored subterranean city?

"Have you heard of the secret White House?" Ken Kellermann of the National Radio Astronomy Observatory once asked me.

"You mean the Greenbrier?" I said.

"Wasn't so secret, was it?" he said.

"Yeah, the locals say they knew about it."

"Makes you wonder, doesn't it?" Kellermann said.

"Like where the *real* secret White House was?"

"You're catching on," he said. "Why would they have done that?"

"It was a decoy for the *real* secret White House?"

"And where do you think that could be?"

"Not in Sugar Grove?"

He shrugged.

Comments like that sent the conspiratorial side of my brain into overdrive. I began hunting for evidence that might support such theories. In reality, I learned with astonishment, the *entire region* was essentially a government-classified nuclear fallout safety area. In the 1980s, the federal government's official crisis relocation plan drew up a scheme for residents of Washington, D.C., to flee two hundred miles to Elkins, West Virginia—the nearest city to Green Bank—in the event of a nuclear strike against the United States. People were to wait out the holocaust in the hills of Appalachia. "In many ways," Graff writes in *Raven Rock,* "after many decades, many billions of dollars, and countless advances in technology, the federal government's basic plan to escape a catastrophe in Washington remains the same today as it was during that first Operation ALERT during the Eisenhower administration: Run away and hide in the Appalachian Mountains across Virginia, Maryland, and West Virginia."

Operation ALERT was an annual defense drill from 1954 to 1961 that tested government and civilian readiness for World War III.

The residents of Pocahontas County were a step ahead of the government's plans. I was already encountering a strong survivalist mindset. "We always laugh that we can't wait for the apocalypse, because we'll be good!" said Jonese. The thinking was that if one could survive here—drinking untreated well water, eating hunted meat and foraged plants, hours from cell service, highways, or a Walmart—then one could survive long after civilization had collapsed in cities. Because a decent-size grocery store was at least an hour away, many people had months of food stocked in their fridge, freezer, and pantry.

The prepper mentality went even deeper. At Trent's, I once met a man named David Warner who told me that he safeguarded about fifty pounds of gold in a nearby cave reachable only by crawling through a hole, crossing a subterranean creek, and ascending a set of dirt stairs to reach an underground lair. His cave held "rifles, shotguns, and semiautomatics," a solar-powered generator, two diesel generators, twenty gallons of gasoline, "just about every tool you can think of," gas masks, and enough food for twenty people to survive for two years—enough for him and his wife, their four children and their spouses, and their ten grandchildren. He saw no need to own a cellphone, but he was looking into buying a radiation suit. "I'm ready for everything but nuclear war," Warner said, "but I hope it doesn't come to that. I have faith in President Trump."

It's easy to laugh off such doomsdayers—at least until an earthquake hits, a pandemic starts, or another disaster strikes. In 2013, the British Royal Academy of Engineering predicted a 50 percent chance that a solar superstorm will occur before 2063, causing mass power outages and disrupting global communications—that is, wiping out cell service and WiFi, turning Earth into a global Quiet Zone. Green Bankers liked to think they'd be ready.

ALL THE SPECULATION and intrigue sent me on something of a conspiracy tour of Appalachia, going inside the decommissioned congressional bunker at the Greenbrier (tickets cost thirty-nine dollars a person) and sneaking a glimpse of Sugar Grove's still-active spy scopes from a ridge outside of town. While I ogled at the highly classified site, Jenna stayed in the car—this was not turning into her best summer vacation.

We drove into the village of Sugar Grove. It was little more than a T-intersection flanked by a Methodist church and Bowers Store, which had a single gas pump in front and a pasture out back. We peeked inside, though we weren't sure the store was open because the lights were off and it was July 4. A dozen rifles were scattered across a glass counter. Ammo was stacked next to vegetable oil. Dusty antiques packed the floor-to-ceiling shelves.

I suddenly realized an old man was sitting behind the counter, practically blending in with the menagerie of taxidermy on the walls and shelves—deer heads, a fox, an owl, a muskrat with ears so disheveled they looked like late-autumn leaves. It was John Bowers, the proprietor, born down the street in 1929. He invited us to look around. We were particularly intrigued by what appeared to be an antique post office facade for sale, but this turned out to be Sugar Grove's actual post office. Bowers's son was the postmaster.

Jenna inquired about a bathroom, and Bowers showed her across the street and into the church, where there was a toilet in the basement. When they returned, I asked Bowers if he also delivered sermons there on Sundays. "I'm not the pastor," he said, "but if you two want to get married, I'll take your fifty dollars and marry you." We laughed awkwardly. We'd been dating for six months. Marriage wasn't yet a topic.

Jenna bought a bag of Gibble's brand Cheese Puffys. The price tag said $3.19 but Bowers charged $3.00. Behind us, a bicyclist in full-on racing gear—black tights, colorful jersey, leather gloves—bought two packs of Twizzlers and a soda, which also came to $3.00. Everything was rounded to the dollar. For the hour or so that we lingered in the store, one other person came inside, and it was Bowers's grandson. The place felt desolate.

After a while, I realized the cyclist had never left. He was still standing out front, casually sipping his soda and occasionally glancing in at us, which felt odd. For one, I didn't see many bicyclists in Appalachia, much less any decked out in colorful spandex. For another, what hard-core cyclist stops for an hour to drink soda? Jenna and I both felt we were being monitored. Perhaps it was paranoia from all the conspiracies we'd been hearing, but it also seemed to make sense, because we stood in the shadow of a major spy facility. Of course, regardless of whether that was actually a government minder, we were probably being *less* watched than ever. Google and Amazon normally monitored our every move through our online habits, tracking our locations throughout the day based on where Jenna used her smartphone or where I logged on to WiFi, divining our intentions from our internet activities and selling that data to advertisers. Such surveillance was less possible due to the Quiet Zone, which, in turn, made it an ideal place to conduct spy work. It was almost comically perverse: the last great radio quiet place in America was being used to monitor the radio noise blaring everywhere else.

Sugar Grove's government operations were divided into two parts: a mountaintop "upper base" that was the site of more than a half dozen dish-shaped antennas, and a valley base that could house up to 450 personnel in a self-contained village that had a gymnasium, swimming pool, bowling alley, tennis courts, water tower, and six-bay fire station. In 2016, the navy shut down operations at the lower base, which was put up for sale. In marketing it, the General Services Administration touted in a glossy brochure that the area was "free from noise and smog" and "you would be hard pressed to find such fresh air and church-like stillness anywhere else."

One interested buyer was a woman named Diane Schou of Green Bank. With several million dollars in financial backing—as she would later tell me—she toured the site on behalf of investors interested in converting Sugar Grove into a haven for "WiFi refugees," people with a mysterious illness called "electromagnetic hypersensitivity." Sugar Grove "would be a great place for people to come and heal," Schou was quoted as saying in the *Wall Street Journal*.

Bowers, for his part, said he wanted Donald Trump to build a casino in Sugar Grove. But based on the president's addiction to his smartphone and social media, it was unlikely that he'd survive long in the Quiet Zone.

Ultimately, the lower base was sold for $4 million to an investor who planned to create a health care campus to deal with the regional opioid epidemic. The upper base would remain an NSA listening post, which had implications for Green Bank. As long as the NSA operated out of Sugar Grove, the astronomers had a powerful ally—one that was showing a growing interest in Green Bank's future.

Amid the cuts in funding from the National Science Foundation, reports leaked in 2017 that the observatory was in discussions for a partnership with the "national security community." Scientists had been speculating about the NSA's interest in Green Bank. Writing in *American Scientist* in 2016, the cosmologist Tony Rothman suggested that "the most reliable way to save the Green Bank Telescope would be to turn it over to the NSA, which would undoubtedly suit many Congressmen." NRAO director Anthony Beasley himself told me that the Department of Defense "could come in and take over the site and not a lot would change."

"DoD is a good option," Beasley said. "We do observations with partners and other agencies in the U.S. We do tests. We don't do surveillance or anything like that. We certainly don't do the stuff that they do in Sugar Grove. But it's not like we don't know those people and haven't done experiments with them."

A sense of the NSA's speculated interest in Green Bank could be drawn from its past usage of the Arecibo Observatory in Puerto Rico. Before being turned over to the National Science Foundation in 1970 to be utilized for radio astronomy, Arecibo was originally built with U.S. Air Force funding to conduct radar studies of Earth's ionosphere. In the '60s, under the guise of conducting a study of lunar temperatures, the NSA used the one-thousand-foot-wide dish to capture Soviet signals as they bounced off the moon. "NSA officials were amazed with the results at Arecibo," James Bamford wrote in *Body of Secrets*. "Just as anticipated, the sensitive Russian signals drifted into space, ricocheted off the moon, and landed, like a ball in the pocket of a pool table, in the Arecibo dish on the other side of the planet."

CHAPTER SIX

"Our Own Log Lady"

STILL IN HIS PAJAMAS, Bob Sheets turned on his kitchen griddle and whisked up a bowl of blueberry pancake batter. I had come to debrief. The Sheetses were becoming my guides to better understanding the Quiet Zone, and my head was spinning. I had come to Green Bank on the presumption that the less connected life was richer—which seemed to be bearing itself out. But I was also staring down a rabbit hole of alien hunters, government spies, and WiFi refugees.

Hovering over the stove, Bob chuckled as I said Green Bank reminded me of the fictional town of Twin Peaks from the 1990s television series. Both Twin Peaks and Green Bank were remote mountain towns with economies built around logging and lumber mills. Both towns had supernatural undertones, with government officials trying to communicate with aliens. The story of *Twin Peaks* was built around a tragic murder, and I was hearing that Pocahontas had its share of killings, some unsolved. Driving around the mountain roads, winding alongside ancient rivers that had eroded gorges into the landscape, with a dark forest on one side and green pasture on the other, I could *hear* the quiet around me, and it sounded like the foreboding, slinky, synthesizer-filled theme song to *Twin Peaks*.

"I think it has to do with the intersection of cultures," Bob said. He poured batter on the griddle. "We've got our mountain culture, scientific culture, and the wave of special people that have 'powers.'" He eyed bubbles forming on the pancakes.

"I haven't seen that show since I was seventeen," his wife, Elaine, chimed in. "I just remember the funky music. And the Log Lady." In *Twin Peaks,* a seemingly crazy woman referred to as the "Log Lady" believes she is the medium for a clairvoyant log that warns of unseen dangers. "We have our own Log Lady," Elaine added.

The Log Lady of the Quiet Zone lived a few miles away and had the mysterious power to "detect" WiFi, cell signals, and other forms of electromagnetic radiation. I was told I might find the Log Lady at church—perhaps the only church in America quiet enough for her to attend.

POCAHONTAS COUNTY HAD about three dozen active churches, approximately one for every 230 people, compared with a national average of one for every 1,000 people. The churches were community gathering places, and I'd found myself relying on them to better understand the area. Most were Methodist or Presbyterian. Some were Baptist and Brethren. Three were Catholic. And Wesley Chapel United Methodist Church, built in 1897 on a hillside about five miles from the observatory, seemed to be the designated church of the so-called electrosensitives.

Wesley Chapel was quiet both by law and by default. In addition to lacking cell service, the church had no plumbing and no telephone, simply because parishioners never bothered with the upgrades. Out back were two outhouses, each with a rickety wooden door, one labeled "Ladies" and the other "Gentlemen." In the front yard was a stone tablet engraved with the Ten Commandments. "Thou shalt not kill," it ordered. For Diane Schou, who felt that her life was physically threatened by the electromagnetic radiation from cell towers, smartphones, and even certain lights, the sixth commandment meant strictly enforcing the rules of the Quiet Zone.

The Sunday that I first entered the white clapboard church, sunlight poured in through four tall windows on either side of the sanctuary—fortunate, given that the lights were kept off for the electrosensitive congregants. I spotted Schou, a heavyset woman with gray pigtails, glasses, a floral-print dress, and a cane. Of eight congregants, she was one of three with electromagnetic hypersensitivity. I scooted into a pew.

Pastor David Fuller, a middle-aged man who wore a suit jacket and crisp dress shirt, nodded hello to me from the pulpit. This was the second of three churches that he preached at every Sunday; two of the three had fewer than a dozen congregants. It would have made sense to consolidate everyone into a single church, but people clung tight to their ancestral houses of worship. One congregant at Wesley Chapel would proudly tell me how his great-grandmother donated an organ to the church way back in 1898. It stood at the front of the sanctuary, never used.

After a few hymns accompanied by an out-of-tune piano, Fuller launched into a sermon about the church's responsibility to evangelize, which was something Schou knew about, as one of the world's most vocal evangelists for EHS. When I introduced myself to her after the service, she was hardly surprised that I'd sought her out. She'd already been interviewed by worldwide media, from *Newsweek*, *National Geographic*, and National Public Radio to Brazil's *O Globo*, Britain's the *Guardian*, and Germany's *Der Spiegel*. Media portrayed her as a canary for the health dangers of wireless connectivity. She had taken it upon herself to write letters to everyone from the chairman of the Federal Communications Commission to the president of the United States about her illness.

Schou was with two friends, Allan Clark and Jennifer Wood, who had also moved to the area because of their shared illness. Wood had long red hair and wore a dark dress. She'd lived in Green Bank for six years and said she walked barefoot around the telescopes daily, electrically "grounding" herself in nature and soaking up the quiet. Clark, chubby and middle-aged, had recently arrived from Missouri on the recommendation of a friend who'd seen the 2016 documentary *Lo and Behold,* in which Wood was interviewed. With a worn look on their faces and a desperation in their eyes, all three of them started telling me how they could *feel* electromagnetic radiation emanating from smartphones, radiating through walls, and assaulting them from all sides, making life unbearable in the outside world. Wood said she personally knew "at least one thousand to two thousand people" with EHS. Many, like her, experienced "pins and needles" in their heads when close to cellphones. In cities, Wood felt like her skin was "frying." She got vertigo around electric stoves and other electronics and developed a metallic taste when around power lines or cell signals.

"Four or five people have died in my arms from this—*literally* died in my arms," Wood said. "I can't stress enough how serious this is. You could be the one tomorrow to get this, and I'm telling you, if you get poisoned like this, your worldview will be 100 percent changed."

For most people, electromagnetic radiation is imperceptible and harmless. It radiates from almost every object, forming a chaotic mess of photons propagating through space at the speed of light. Strong doses *can* be dangerous. The Food and Drug Administration sets a limit on how much radiation can leak from a microwave oven, while also recommending that users not stand directly against the appliance. A single hospital CT scan is considered safe, but the American College of Radiology recommends limiting your lifetime total to twenty-five chest CTs because of the rising risk of cancer. On a more extreme level, the U.S. military's arsenal includes a microwave-radiation weapon that makes people feel like their skin is burning. For the electrosensitives, almost any dosage of human-made electromagnetic radiation could make them feel ill.

Schou believed that she was the first electrosensitive to arrive in Green Bank, in 2007. She estimated that more than one hundred people with EHS had since moved to the area from around the world, driving up housing prices and creating a market for "quiet" real estate. The *Pocahontas Times* regularly listed new properties with a "great location in the Quiet Zone," and the electrosensitives had come to make up about 10 percent of all home sales for Marlinton-based Red Oak Realty—on pace for about two hundred home sales in the decade after 2010. Every week, Red Oak Realty got two to five phone calls from electrosensitives looking to buy property. They often brought their own handheld field-strength meters to test noise levels around houses.

Schou invited me to talk more at her home a mile down the road. She moved slowly with the aid of a cane. We loaded into our cars and I followed behind her maroon Subaru Forester, past rolling farmland and forest, to her large brick house. A camper van with an Iowa license plate was parked outside. Along the driveway was a tiny red cabin. It was occupied at the moment by a woman named Julie with EHS. Two other sensitives were camping in Schou's backyard. Wood had also stayed here when she first arrived in 2011. Other people with EHS were known to simply park at the observatory and sleep in their cars until they were asked to leave.

Schou also owned a house across the street that she rented to families with sensitivity. She'd purchased it because the previous owner had WiFi and she wanted quieter neighbors. Several miles away, she oversaw a fourteen-acre wooded property, which was owned by a nonprofit 501(c)(3) she'd founded called Wave Analysis Verification Research (WAVR). It had two cabins without water or electricity. Schou hoped to develop WAVR into a full-service "resort" to host the sensitives who called her daily from around the world asking for help.

While not recognized as a medical disorder in the United States, EHS has some high-profile advocates. Jolie Jones, daughter of the renowned jazz musician Quincy Jones, claims to have developed EHS around 2008. Gro Harlem Brundtland, the former prime minister of Norway and former director-general of the World Health Organization, also believes she suffers from EHS. Jack Dorsey, the CEO of Twitter, reportedly owns a $5,499 electromagnetic radiation-blocking tent and feels better when he's "not getting hit by all the EMF energy." Jill Stein, the Green Party's candidate for U.S. president in 2016, said on the campaign trail that she opposed WiFi in schools on the grounds that "we should not be subjecting kids' brains especially to that."

EHS has gained wider prominence since the 2015 premiere of the AMC series *Better Call Saul,* which featured a character named Chuck McGill who wrapped himself in a silver blanket to protect himself from electromagnetic radiation and lived in a house disconnected from the electric grid. Schou called the show a "very good educational tool" based on the two episodes she'd seen. She apparently didn't know McGill was eventually revealed to be a hypochondriac.

Inside Schou's living room, a five-column bookcase held six rows of shelves overflowing with folders, binders, and stacks of paper. It was a library devoted to EHS. She pulled down a booklet titled

Electromagnetic Sensitivity and Electromagnetic Hypersensitivity: A Summary. It listed nearly one hundred potential symptoms of EHS, including earaches, tinnitus, high blood pressure, low blood pressure, confusion, anxiety, depression, fatigue, memory loss, muscle weakness, restless legs, shiny eyes, dry eyes, allergies, hair loss, and pain pretty much anywhere in the body. It seemed like *anything* could be attributed to EHS. As mystifying as it all sounded, Schou claimed to speak with a degree of authority, as she had a doctorate in industrial technology from the University of Northern Iowa (which the university later confirmed for me). She regularly signed letters as "Diane Schou, Ph.D."

Schou's most telling symptoms of EHS had been hair loss, rashes, and headaches, which she began to suffer in 2002. "And I rarely had headaches [before 2002]," she said, "unless I ate ice cream really fast." At the time, she was living on a farm outside Cedar Falls, Iowa, with her husband, Bert, and their son, Paul. The company UScellular had recently built a cell tower near their home and, according to Schou, whenever she neared it she felt a "sledgehammer" of pain in her head. Her husband and son weren't bothered, so Schou concluded she was especially sensitive. She petitioned the U.S. government to investigate the tower, to no avail. To get away from the pain, she began sleeping in a homemade Faraday cage—essentially a box covered in wire mesh that helped block out electromagnetic radiation (not unlike that tent where Jack Dorsey takes refuge). Schou's cage was just big enough for a twin bed. Bert delivered her meals there. "I would have to leave occasionally so I could go to the bathroom or take a shower," she said.

Unless she wanted to live that way forever, Schou believed she had to leave Cedar Falls. She traveled across the country in the family Winnebago, logging thousands of miles as she searched for a place where she might feel free from pain. She and Bert visited an

EHS sanctuary in Snowflake, Arizona, one of many communities that have cropped up around the world for people with unknown illnesses. But the high desert altitude didn't agree with Bert. She visited a remote island in Nicaragua where she felt better, but the language barrier was too challenging. She eventually met a forest ranger who mentioned a place in the Appalachian Mountains where cell signals were restricted. To Green Bank, Schou brought a new disease that most people had never heard of—though in time she would convince some locals that they, too, were sensitive. She purchased a burial plot in the cemetery behind Wesley Chapel, intending to live out her days in the Quietest Town in America.

While I was having a hard time understanding Schou's illness, I wondered if her reaction to cell service and WiFi was an intense manifestation of the kind of tech overload that we all experience at one time or another. Although she was an extremist, there was something very human about her search for quiet, as if hers was part of a bigger quest that spanned generations, civilizations, and belief systems.

CHAPTER SEVEN

"A Powerful Thing"

NEARLY THREE THOUSAND YEARS AGO, the Jewish prophet Elijah hid in a cave and heard God not in a passing windstorm, earthquake, or fire, but in "a still small voice." Thousands of miles away around 500 BCE, Siddhārtha Gautama reached enlightenment while quietly meditating under a tree. The ancient Greek philosopher Epicurus espoused "a quiet life." Jesus emerged from a forty-day solo desert retreat to preach the Gospel. Hinduism considers outer and inner silence, or *mouna*, a sacred way of purifying the mind. "All of humanity's problems stem from man's inability to sit quietly in a room alone," concluded the seventeenth-century philosopher Blaise Pascal. The nineteenth-century Danish theologian Søren Kierkegaard prescribed quiet to remedy "all the ills of the world." The twentieth-century monk Thomas Merton embraced monastic silence as a way of coming closer to God.

From what I could tell, Green Bank had what many gurus sought across the ages. It was also a *real* place—a living, breathing, pulsing community—unlike the contrived getaways at increasingly popular "tech-free" retreats that provided temporary reprieves from the digital overload. Around the time that Jenna and I first visited Green Bank, we had also stopped into a silent retreat center in rural

Massachusetts where people paid for the privilege of locking their smartphones inside a large metal safe while they engaged in meditation. I'd read about the retreat in a *New York* article titled "I Used to Be a Human Being," which described the writer Andrew Sullivan's ten-day digital detox. According to Sullivan, for a decade he had lived inside "a constant cacophonous crowd of words and images, sounds and ideas, emotions and tirades—a wind tunnel of deafening, deadening noise." So he decamped to the Insight Meditation Society's silent retreat in the New England woods. Outside of such getaways, the only remaining phone-free "safe spaces" were the shower and the desert festival Burning Man, Sullivan jested. The iPhone was waterproofed in 2018 and Burning Man got cell service in 2016. Even airplanes, once high-flying zones of disconnection, now offer WiFi. What quiet remains?

This loss of radio quiet has coincided with a loss of audible quiet. In 2000, the director of the U.S. National Park Service passed an ordinance on "soundscape preservation and noise management" that called for parks to document and work to preserve natural sounds. The directive expired in 2004. Three years later, when the iPhone debuted, *Science* reported that human-made noise pollution was "pervasive" in America's protected areas. An acoustic ecologist named Gordon Hempton today believes that only a dozen places remain in the United States where a person can hear no man-made sounds for fifteen minutes. Three of the remaining places are the Hoh Rain Forest in Washington, Boundary Waters Canoe Area in Minnesota, and Haleakalā National Park in Hawaii. Hempton has refused to reveal any more locations in order to protect their silence. Once a given aspect of nature, quiet is facing extinction.

I've witnessed the loss firsthand. At the northern terminus of the Appalachian Trail, Maine's Baxter State Park is renowned for its strict wilderness policy—you can't even find a water spigot inside the park. Home of Mount Katahdin, the state's highest peak, the park bans dogs, alcohol, and "the use of electronic devices in any way that impairs the enjoyment of the park by others." That directive was meant to include cellphones. "Something as simple as someone pulls out their cellphone on top of Katahdin and makes a call to say, 'You won't believe where I'm calling from right now'—that takes away from someone else's wilderness experience," a former Baxter chief ranger said in a 2018 news article. When Jenna and I climbed to the summit of Katahdin that year, however, we heard someone talking on their smartphone. Rangers had essentially abandoned the no-phones rule because of pushback from visitors. An otherwise pristine wilderness was marred by a blinking cell tower on the horizon.

The fight for quiet has seeped into popular culture. The horror flick *A Quiet Place,* about a family trying to avoid the detection of monsters with hypersensitive hearing, is really a fable about Western culture's "war with noise," according to one media critic. We need "quiet cars" on the train and reminders at the start of every movie to put our cellphones on quiet mode. Even smartphone developers have been capitalizing on the search for quiet, with meditation apps such as Calm and Headspace valued in the hundreds of millions of dollars. The company Yondr does a brisk business selling cellphone pouches that prevent people from using their devices at places like schools, courts, and concert halls. Some musicians and theatrical performers have taken matters into their own hands, stopping mid-show to ask people to turn off their devices. In 2015, the actor Patti LuPone made headlines for grabbing a phone from a theatergoer.

More than annoying, human-made noise has been shown to increase the risk of heart attack, stroke, diabetes, and even cancer. Bison-related injuries have soared in Yellowstone because of the selfie-spurred urge to get dangerously close to the wild animal.

Between 2011 and 2017, according to one study, 259 people worldwide died while taking selfies, primarily from drowning, falling, and getting into transportation accidents. America is suffering from a "national attention deficit" as a result of our devices, according to the University of Wisconsin neuroscientist Richard Davidson, founder of the Center for Healthy Minds. For years, internet addiction has been classified as an official disorder in South Korea and China, where hundreds of government-run health clinics treat addicts. In 2018, the World Health Organization added "gaming disorder" to its *International Classification of Diseases.* We might as well add "smartphone disorder" and "social media disorder." A 2018 study found that, despite the fact that people are measurably happier after giving up Facebook, you'd still have to pay the average user $1,000 to convince them to deactivate their account for a year—meaning that people are choosing to be unhappy, such is the controlling nature of social media, enabled through smartphones and constant connectivity.

Even God wants people to reconsider their smartphone habits. Pope Francis has called on followers to "free yourself from the addiction to mobile phones." Since 2009, the Jewish nonprofit Reboot has organized an annual National Day of Unplugging "as a way to bring balance to the increasingly fast-paced way of life and reclaim time to connect with family, friends, and our communities." In Green Bank, a Mennonite leader suggested to me that the lack of cell service made it easier to be Christlike. His internet was "strongly filtered," and all church members set limits on screen time. No one had televisions. I asked if he indulged in Netflix. "Um, that's movie streaming?" he said of a company with more than 150 million subscribers. "We don't really do that."

And there are plenty of non-religious reasons to take a break from screens. Smartphones have been blamed for a rise in adolescent depression, anxiety, sleep deprivation, and suicide. It's so much easier to feel isolated and left out when you're constantly seeing Instagram posts of everybody else having fun. It's also easier to not go out when your life is online. The psychologist Jean Twenge of San Diego State University has repeatedly warned about a generation lost to smartphones. "It's not an exaggeration to describe iGen as being on the brink of the worst mental-health crisis in decades," Twenge wrote in the *Atlantic* in 2017. "Much of this deterioration can be traced to their phones." Teens are driving less, dating less, and having sex less, while still being less likely to get enough sleep.

Falling fertility rates worldwide have been blamed in part on our preoccupation with smartphones, video games, and social media. According to a 2012 study from the psychologists Netta Weinstein and Andrew Przybylski, who later became director of research for the Oxford Internet Institute, "the mere presence of mobile phones inhibited the development of interpersonal closeness and trust, and reduced the extent to which individuals felt empathy and understanding from their partners." Studies have shown that people who text heavily have more difficulty falling asleep or psychologically disengaging from job-related stress, and that smartphones promote lazy thinking and are a major cause of car accidents. Matt Richtel's book *A Deadly Wandering* powerfully laid out the consequences of mixing cellphones with cars: talking on a phone while driving is equivalent in crash risk to driving drunk, and texting while driving is even more dangerous. Between 2015 and 2016, U.S. traffic fatalities surged 14 percent, the biggest spike in more than half a century, which the Governors Highway Safety Association blamed on distracted driving due to smartphones. Pedestrian deaths hit a nearly thirty-year high in 2018, also credited to smartphones.

Wouldn't there be fewer deaths and crashes if it was impossible to drive and text at the same time? Wouldn't all of us sleep better if we lived in a place without constant connectivity? Wouldn't our lives be richer and our communities stronger if we were not always online? And if all these benefits of a less digitized life were true, wouldn't Green Bank and the surrounding Quiet Zone be a kind of utopia?

IN 2016, an Italian researcher named Goffredo Colini visited Pocahontas to measure whether people were happier in a place largely lacking wireless connectivity. As part of his master's project for Salesian Pontifical University in Rome, Colini conducted surveys of 220 people in Pocahontas and the nearby city of Lewisburg. (Such a study would have been impossible in Italy, Colini told me, because of blanket cell service.) He determined that residents of the Quiet Zone experienced measurably lower stress and anxiety. He also found himself charmed by Pocahontas, despite initial alarm over the number of guns he saw. People went out of their way to answer his questions. He was invited to a bachelor party in the forest where he drank moonshine (presumably better than the batch I tasted). A random family gave him a ride to Washington, D.C., so he could catch his flight home. Everyone was amiable and willing to interact. "It was the first time in my life that people were focused on just how they can help you," Colini told me.

Community seemed enhanced by the quiet. The lack of cell service created a greater sense of self-reliance, but also of reciprocity. "This winter, I don't know when my ass is going to be in a ditch and who's going to come along and pull me out," as Wesley Sizemore said to me. "And if *their* ass is in the ditch, I'll pull them out as well. That's an Appalachian thing. Get rid of cellphones and Facebook and Twitter, I think everybody would revert back to that."

Without cell service, locals found other ways to communicate.

Pocahontas County's local HAM radio club routinely acted as a kind of vigilante rescue service. With more than one hundred licensed amateur radio operators, the club claimed to have the highest per capita number of HAM radio operators on the East Coast—as well as to be the county's most "connected" organization. Club member Rudy Marrujo had come across vehicles broken down on Route 28, stranded on Cheat Mountain, and with a flat tire at Buffalo Lake, and in each case he'd called for help using his car's HAM radio. The club said it operated on frequencies rarely used by astronomers.

Because of the lack of cell service and notoriously patchy landlines, sometimes the only way of communicating was via HAM radio or Allegheny Mountain Radio, the county's sole radio station. About ten miles south of Green Bank, a 180-foot-tall radio antenna transmitted at 5,000 watts, about one-tenth of the standard output for radio towers, in keeping with the Quiet Zone restrictions. (Because the station broadcasted at 1370 kilohertz, the transmission was negligible to the observatory. Astronomers didn't observe frequencies below 100 megahertz, or FM 100 on the car radio.)

In 2012, when a severe windstorm knocked out the county's power grid and telephone lines for more than a week, the HAM club radioed for help on behalf of the injured and stranded. The sheriff put out a message on Allegheny Mountain Radio requesting assistance from the observatory, which was running on diesel generators. The facility turned itself into an emergency shelter, providing the community with food, water, and washing facilities. The *Pocahontas Times* praised the observatory as "a beacon of hope during this dark time."

It was not the first instance the observatory acted as an emergency service provider. In February 2010, when a navy MH-60 Knighthawk helicopter with seventeen people aboard crashed in the nearby forest, the observatory turned one of its offices into a makeshift command center for the rescue. A husband-and-wife team of bear hunters went out on four-wheelers. A fire chief attempted to bulldoze his way to the crash site through eight-foot-high snowdrifts on old logging roads. By morning, a team of snowmobiles and grooming machines from Snowshoe Mountain Resort evacuated everyone. The governor held an appreciation ceremony for the observatory's coordination efforts.

Without the ability to call for help on a cellphone, one had to trust in the help of strangers. A teacher told me she was once driving over a mountain when she hydroplaned, spun in a circle, hit a tree, and smashed her windshield. She'd had an iPhone, but there was no service. After forty-five minutes, a man driving by with his Chihuahua offered to give her a ride to her parents' house. Necessity bred kindness. "If you come upon a tree in the road, don't panic, because someone will be coming along in a few minutes with a chain saw," said Jaynell Graham, editor of the *Pocahontas Times,* a weekly newspaper that in itself reflected that quiet community feeling.

Nicknamed the *Poky Times* because of its pace in getting around to the news with just two full-time reporters, the paper was founded in 1883 and based in the county seat of Marlinton. Graham, who'd been editor since 2012, focused on upbeat stories about schools, family reunions, and history, boasting that the *Times* was one of the few regional papers that didn't put police logs on the front page. "I try to focus on the positive," she told me. "Sometimes it looks like a school newspaper. Kids killing their first deer. Honor roll. My thinking is, the more positive coverage you give to these kids when they're young, the less likely you are to see them in the magistrate court." Print circulation was 4,300, plus another 400 e-subscribers—not bad for a county with around 8,200 residents.

Aside from running the *Times,* Graham raised a small herd of cattle, played the organ in her church, put together the services' liturgy and prayers, typed up the weekly bulletin, and was the county's Republican ballot commissioner. She exemplified how, in a small community, people wore multiple hats and knew everybody. I began regularly swinging by the *Times*'s office to get the latest news from Graham.

I also would have benefited from getting a HAM radio license to stay in the loop. So many HAM operators were monitoring their scanners that when law enforcement radioed in an emergency, the club itself sometimes responded to the accident. HAM operators were also known to help relay information to the police dispatch in Marlinton when officers were out of two-way radio contact. Occasionally, law enforcement simply asked to use residents' landline telephones. Gossip spread fast in a place where people talked on open channels. In general, it seemed people just liked to talk. Outside the sheriff's office in Marlinton, a spray-painted sign read "No Talking to Inmates," apparently because passersby had a tendency to chat with prisoners through the iron-clad windows. In the past, inmates had been kept company by the jail keeper's children, one of whom told me she used to play cards with the prisoners.

"It's a whole different set of rules here," said David Jonese, the former sheriff. "Everyone thinks it's Mayberry—you know, *The Andy Griffith Show*. They think your worst thing is a cat in a tree. It's nothing like that. But I will tell you, my very first call on the job was a loose horse on the golf course."

CONTEMPORARY LANGUAGE equates disconnectivity with death. We say "the line went dead" or we drove through a "dead spot," and when reconnected we're "back live," with the implication that life only happens in places with connectivity.

"Through the lens of our techno-normative values, the cultural significance of a dead zone is a mistake, an error to be corrected, a gap in service or infrastructure that must be addressed," Nathanael Bassett of the University of Illinois in Chicago would write in his Ph.D. dissertation, titled "Dead Zones: A Phenomenology of Disconnection." His research led him to Green Bank, where he discovered not a lifeless community but a place where people were "outwardly focused" and neither isolated nor alone. The Quiet Zone helped residents take control of when they chose to be online; it provided "freedom from interference, and freedom to explore other connections," Bassett concluded.

That aspect of the Quiet Zone was invaluable to Sarah Riley, executive director of High Rocks Educational Corporation, which operated a two-hundred-acre leadership camp for teenage girls at the southern end of Pocahontas County. The Quiet Zone created space for youths to open up.

"It used to be scary for them to come to a place where they didn't have any friends," Riley told me when I visited the camp that summer. "Now it's so scary for them to be in a place where they're not going to be connected to the world through their phones and have to give up that lifeline. For young people that are able to come to our programs and experience that, it's cited as a powerful thing."

Everyone at the camp—called High Rocks Academy—attended on scholarship, and all had to apply to be admitted. Among the incoming group of two hundred campers in June 2017, one had written in her application, "The majority of kids today don't want to go to these camps since there is no internet or cell service. For the ones that do go, this will be a good chance to be outside more." Another wrote, "I will unplug from the internet and social media and I will become a part of something bigger." Another said, "I've been attached to the WiFi for too long." Even if these young women were terrified of being disconnected, they each seemed to know it would be worthwhile.

More than five hundred youths had gone through High Rocks' summer program since it was founded in 1996, and many came from low-income homes where high school was the highest level of education. Hundreds more young men and women had received mentoring, tutoring, and college prep. The organization had active partnerships with Appalshop, a well-known arts and education center in Kentucky, and Highlander, a Tennessee-based social justice leadership training school, to develop youth leadership, which Riley called a "cornerstone" for building sustainable and inclusive communities in central Appalachia. High Rocks was also partnering with the Green Bank Observatory to boost the state's science, technology, engineering, and math (STEM) education, with a focus on rural first-generation college students, funded by a $7 million grant from the National Science Foundation.

Coming from a family of educators, Riley was well familiar with the unique challenges facing youths from Appalachia. Her sister was the county math coach. Their mother, Susan Burt, was a native Michigander who had attended Pomona College and led the county's public schools' gifted program for twenty years. Their father, Gibbs Kinderman, was a Californian who had attended Harvard and eventually settled in Pocahontas, where he helped found Allegheny Mountain Radio and was a longtime board of education member. Riley had also attended Harvard, which was a pivotal moment in her own education, though for unexpected reasons. Arriving in Cambridge in 1993, she'd gone from a home without a TV to an internet-connected dorm room. She found herself struggling to understand the academic language of her classmates and facing derogatory jokes playing on Appalachian stereotypes. But by her junior year, things clicked. She could write a thesis. She could deliver an argument. She was learning the language of power. And she was realizing that her peers back in West Virginia were as smart and capable as most anyone at Harvard. It fueled her desire to empower youths from West Virginia. When she graduated in 1997, she returned to Pocahontas to work at High Rocks alongside her mother, who had recently founded the organization on the idea that every child could perform at gifted levels if given the opportunities to thrive.

Riley appreciated how the Quiet Zone enhanced the High Rocks experience, even if the camp's location in Pocahontas County complicated operations. One of those complications was slow internet, as the county seemed unable to attract significant investment in broadband. Riley considered fast internet indispensable for managing the camp, so she'd had a dedicated fiber-optic line installed through a federal subsidy for educational facilities. On review, however, the Federal Communications Commission had determined that High Rocks did not qualify for the subsidy, putting Riley into a protracted dispute with the FCC and internet provider Frontier Communications over a $5,000 monthly bill that soon added up to six figures.

Without that fiber-optic line, Riley's internet was so slow that she couldn't download emails on snowy, windy, or rainy days. Such was a general hurdle to education in Pocahontas. Her husband, Joseph Riley, was the principal of Pocahontas County High School, and he'd found it necessary to always warn teaching applicants that the area lacked "modern conveniences" like fast internet, fast food, and cell service. With one candidate, he recalled, "We said there's no cell service, no McDonald's, and no Walmart—and he was out!"

I would meet one teacher who never got the warning. It took Teresa Mullen a week before she learned why her cellphone wasn't working, after moving to Pocahontas in 2010 from Pittsburgh to teach culinary arts. While filling her car's gas tank, she asked a stranger tending his pump, "What's your phone provider?" The man didn't know what she meant. "You know," Mullen continued, "your *cellphone* provider, is it AT&T, Sprint, or what?" The man responded, "Oh, honey, we ain't got none of that around here."

Mullen had moved from a city where she could order pizza and Chinese after midnight to a county where the biggest retailer was Dollar General and the only fast food was Dairy Queen. She considered quitting, but she'd already sunk her cash into relocation expenses, so she hunkered down. She got a landline phone, her first since moving out of her parents' house nearly a decade earlier. She got a West Virginia driver's license and upgraded to a Subaru Crosstrek for the mountain roads. She became a regular participant in the county's annual Roadkill Cook-Off, where teams competed to create the tastiest dish from what could potentially be roadkill: turtle, possum, squirrel, rabbit, mink. For one year's cook-off, Mullen's students cooked bear meat burgers. The bear had been shot that same morning by a young woman in the class. "It was still warm when it got here," said Mullen. "There was still hair on it."

HIGH ROCKS FACED another complication, one that I felt uncomfortable even bringing up with Riley: a nearby neo-Nazi organization. Not just any obnoxious group of racists, this had been the longtime headquarters of the National Alliance, once considered a terrorist threat by the FBI and called "the most dangerous and best organized neo-Nazi formation in America" by the Southern Poverty Law Center, a nonprofit civil rights watchdog. While the group had faded in the fifteen years since the death of its notorious founder, William Luther Pierce, it was still expounding its racist philosophy just up the road from the girls' camp.

Riley said parents had in the past expressed concern over an infamous neo-Nazi compound being a mile away through the woods, but less so with the organization's decline. Since I was nearby, I figured I should see for myself what remained of a group associated with violent crimes, murders, burglaries, and plots to overthrow the federal government.

CHAPTER EIGHT

"Back to the Land"

I TURNED OFF ROUTE 39 and onto Boyd Thompson Road, which hardly even seemed like a road—just a dusty, narrow strip of hard-packed dirt cutting uphill between forest and fields. A half mile up, I came to a blue trailer where a gaunt, toothless woman stood outside with two Rottweilers. She waved for me to stop.

I shifted into park and got out of the car. The Rottweilers ran up and sniffed my leg.

"What are you looking for?" the woman yelled, not so welcomingly. A burlap sack was slung over her shoulder. She held a garden hoe.

I was trying to find the National Alliance, despite warnings to stay away. "People like that are liable to shoot you," a county official had told me. The organization's founder hadn't thought highly of the media, as made clear in his novel *The Turner Diaries,* about an underground organization of white nationalists that kills journalists, Jews, and non-whites and overthrows the government by flying a nuclear-armed crop duster into the Pentagon. "One day we will have a truly American press in this country," William Pierce wrote in *The Turner Diaries,* "but a lot of editors' throats will have to be cut first."

The gaunt woman said she was heading toward the compound and I could follow her, though she added that I was unlikely to be allowed inside. She walked ahead with her dogs as I rolled behind in my car.

After a half mile, we came to a metal gate. A burly man with a bushy blond beard and tattooed forearms roared up to the other side on a four-wheeler, a rifle strapped to the front. He wore a T-shirt, cargo shorts, and work boots. The woman disappeared into the woods with her garden hoe, her dogs running ahead.

My heart was pounding. I'd never met a neo-Nazi before—at least, not that I was aware of—and my mind conjured violent images from the movie *American History X*.

"Who do you work for?" the man growled from the other side of the gate. "Do you have any contact with the Southern Poverty Law Center? You don't do any subcontracting for G4S or Blackwater, do you? You don't have a camera on you?"

When I squeaked that I was a journalist researching the National Radio Quiet Zone, the man's face brightened. He said his name was David Pringle and he was "chief of staff" at the compound. He lived there alone with his wife, though a few more people associated with the organization also lived nearby. He added that the National Alliance's headquarters had in recent years relocated to Tennessee, near the home of the new chairman, a former U.S. Army Special Forces captain named William White Williams.

"We call this our West Virginia campus," Pringle said. "We're rehabbing the place after a decade of it being moribund." He motioned to the tall brush around him. He'd just been weed-whacking and his legs were covered in grass clippings. I asked if I could see the work he was doing. First, he wanted to see my I.D., so I passed him my driver's license. I wondered if he was gauging the potential Jewishness of my surname. After a moment he started undoing the chain around the gate and swung it open, motioning for me to pull my car inside.

"Ever rode an ATV?" Pringle asked. I nodded. (There was that one time as a kid . . .) He motioned for me to hop on his ride—an odd proposal, to be sure, but I didn't feel like I could dictate the terms of my tour. Pringle climbed on and I saddled up behind him in my khaki shorts and button-down shirt. He was drenched in sweat. "I've been working all day so I probably have a little odor," he warned a moment too late.

Pringle kicked the four-wheeler into gear. I figured I was only being allowed entry because I was white, and perhaps also because he hoped to somehow co-opt my pen and paper for media coverage. I wanted to see inside, but I didn't want to be unwittingly used to advertise his ideas.

This had been home to the National Alliance since 1984, when Pierce purchased the 346-acre mountainside for $95,000 and built a combination country retreat, business headquarters, and militia base from which to inspire a "white awakening." The land title of sixty acres was held by the Cosmotheist Community Church, which was founded by Pierce and described by the Anti-Defamation League as "a racist religion that stresses the superiority of the white race and the unity of the white race with nature." Calling it a church got Pierce a state tax exemption as well as an argument for protection from religious "discrimination." It also allowed him to perform weddings, including his own to several European mail-order brides. In 2001, he officiated the Cosmotheist wedding ceremony of Billy Roper, a top staffer at the time.

We drove up to a two-story, warehouse-like building that was adorned with a giant, upside-down peace sign. Pringle called the symbol a life rune, a badge of Cosmotheism. It was also a symbol of Nazi Germany. I followed him inside the double doors. A table by the door displayed framed photos of Pierce, a kind of shrine to the man. A magazine rack held old issues of *National Vanguard,* the organization's glossy magazine, as if ready for people to come and peruse the hate literature. Tables and other furniture were splayed haphazardly, and the thin carpeting was stained and fraying. The air smelled musty. This is what remained of the organization's business offices, which once had nearly two dozen full-time staffers. Behind the building, Pringle showed me a large cement cistern with several large cracks. Repairing it was one of his summer projects. The cracks had been patched several times, he said, but the cistern needed a total overhaul to be fixed right. It seemed he was making an analogy for the entire organization.

I asked why Pierce had chosen to base the National Alliance here, of all places. Pringle began talking about the 1921 Battle of Blair Mountain, one of the largest armed insurrections in American history, when more than ten thousand coal miners marched on Charleston to demand the right to unionize from a state government controlled by the mining industry. In a convoluted way, that history would have appealed to Pierce, who dreamed of overthrowing the federal government and installing a pro-white regime. The county's predominantly white population and Confederate roots were also selling points. In the Civil War, 550 men from Pocahontas County fought with the Confederate army, while only 84 joined the Union. At the 1863 Battle of Droop Mountain in southern Pocahontas, in which the Union pushed back the Confederates and essentially took control of West Virginia, the South was vastly outnumbered, but during an annual reenactment more locals always wanted to dress as Confederates, underscoring the area's long-standing alignment with Dixieland.

Pierce was also attracted to the isolation in nature. "I wanted to live on land on which I could hunt my meat and grow my own fruit and vegetables if I had to," he wrote in *National Vanguard*. A 1986 issue of the magazine featured a story titled "Back to the Land: Our Source of Spiritual Health." In the Quiet Zone, Pierce and the neo-Nazis had essentially gone back to nature, joining a movement that had already been thriving in the area for several decades.

SOON AFTER the astronomers arrived in Pocahontas, but before the electrosensitives and neo-Nazis, hippies, and back-to-the-landers had flooded into the county in search of a quieter way of living, so much so that Pocahontas saw its population rise about 15 percent in the 1970s after decades of decline. Hundreds of thousands of urban-to-rural émigrés nationwide were moving back to the land on homesteads, and thousands were coming to West Virginia, as chronicled in the 2014 book *Hippie Homesteaders: Arts, Crafts, Music, and Living on the Land in West Virginia*. Highlighting the bohemian influx, as many as ten thousand flower children flocked to the surrounding Monongahela National Forest in 1980 for a back-to-nature festival called the Rainbow Gathering. As if to put an exclamation mark on the peacenik powwow, a long-haired hippie doctor named Hunter "Patch" Adams arrived in Pocahontas the same year, purchasing 310 acres with the stated mission of opening a free hospital.

Young people were coming from Atlanta, Boston, Chicago, Detroit, and as far away as California in search of unspoiled forestland, low population density, and distance from cities—characteristics enshrined by the National Radio Quiet Zone. Among the first to arrive were Carl and Barbara Hille, caving enthusiasts who moved to the Green Bank area in 1968 from Maryland. They were married by David Rittenhouse. When their son was born, two women from the observatory gifted them a pig. The Hilles' farmhouse became a gathering place for cavers and hippies, hosting about 150 visitors a year.

"They were a different class of people," Eugene Simmons, the county's longtime prosecuting attorney, said of the newcomers. "They were land-lovers that smoked a little grass, were more relaxed than we were."

"They had this idea of what they wanted out of life," said Jerry Dale, who served twelve years as county sheriff and one term as magistrate judge through the '80s and '90s. "Run naked, raise their children how they wanted, grow organic food and a little weed. And you know what? Have at it."

To Dale's point, in 1971 a hippie originally from New York named Laurie Cameron was one of the first people—if not *the* first person—in West Virginia prosecuted for growing marijuana, which got him a year's probation. Cameron didn't stop growing marijuana in Pocahontas, he just got more discreet about it. Not that all the back-to-the-landers did drugs. Jason Bauserman, who moved to the Green Bank area in 1971, said he and his wife were simply called "hippies" because they were outsiders with long hair, even though they were churchgoing Christians who never touched alcohol.

Many of these newcomers settled at the southern end of the county in the mountain hollow of Lobelia, where one road became known as Hippie Hollow. And they were still living there when I showed up in 2017 to visit a new educational nonprofit called the Yew Mountain Center, created by some of the original back-to-the-landers in the area. Named after the surrounding Yew Mountains, the center aimed to be a mix of nature preserve and teaching facility for outsiders to disconnect from modern life and reconnect with the forest primeval—essentially, it was quiet camp.

To get to Yew, I drove up Droop Mountain and down swerving switchbacks so gnarly that one section of road was called Hell's Gate. A quarter-mile-long gravel driveway led up to Yew's massive lodge. I spotted several young women milling around the lawn with infants. A naked baby crawled through the grass. A handsome, bearded man waved to me from a barn. I felt like I'd wandered into a forest oasis and half expected a winged fairy to prance out of the woods and alight a flower wreath around my neck.

I was greeted by Yew's director, a slender lady named Erica Marks who had a daughter on her hip and two more clinging to her legs. All three children had been home births facilitated by a midwife named Danette Condon, who had moved to the area in 1980 and still lived off the grid nearby. "With my first one, I couldn't get in touch with Danette, so we started it ourselves," Marks told me. She described the home births as intimate, in a "quiet, dark place where you're not stressed out." The alternative was driving an hour to the nearest birthing hospital.

A Virginia native, Marks first came to Pocahontas in 2007 to work at High Rocks. When this five-hundred-acre property in Lobelia came up for sale in 2015, she was able to find an out-of-state benefactor—the friend of a friend of a friend—who was interested in safeguarding areas for water conservation and endangered species. "The benefactor was interested in Yew's proximity to the Quiet Zone," Marks told me. It was "remoteness cred."

A dozen people from as far away as Charlottesville had just visited Yew for a birding program, finding sixty-three unique species around the property. Marks was organizing a nature series about the area's geography and wildlife, including how to use the forest sustainably through activities such as foraging. As an example, she pointed to a jar filled with slivers of slippery elm bark. She called it "tree jerky" and said it could be ground into baby food or eaten for medicinal qualities. She took a piece and offered me some. It tasted slightly minty, with the texture of gum wrapped in paper. Marks chewed happily.

In the lodge's kitchen, we found an older woman named Ginger Must chopping vegetables. Fast-talking and spritely, she volunteered regularly at the center. Her husband was Yew's board chairman and a longtime county doctor. He was also among the earliest back-to-the-landers in Lobelia. If I wanted, Ginger said, I could swing by their off-grid home to learn more about the area's history.

But first, Marks was inviting me to join her family for a swim in Yew's spring-fed pond. I didn't want to impose, but I also knew I'd be a fool to decline the offer on a hot summer day. I followed Marks, her husband, and their daughters to the pond behind the lodge. Taking a cue from Marks's husband, I stripped to my underwear, stepped onto a wooden dock, and cannonballed into the fish-filled pond surrounded by slippery elms.

WET UNDERWEAR SOAKING through my shorts, I knocked on the Musts' door, a few miles deeper into Lobelia, toward a remote mountaintop clearing known as Briery Knob. Ginger welcomed me in and offered me a salad. Her husband, Bob, was tall and lanky with short white hair and a fuzzy beard. He said he couldn't sit for long because he had a list of to-dos that he was trying to check off late in the day. He let me pepper him with questions while he painted a closet.

Born in California and raised on air force bases in the United States and England, Bob had been drafted into the army in the late 1960s, serving for two years. When he got out, he wanted to move to the mountains, so he hitchhiked around the country, through the Rockies, Smokies, Adirondacks, and Appalachians. Back home in Georgia, he went to the Emory University library and researched the least populated, most undeveloped mountain land on the East Coast with temperate weather and adequate rainfall for growing food, which pointed him toward Pocahontas. In 1973, he hitchhiked into town with a ponytail and a couple thousand dollars. He soon heard about a mountain for sale in Lobelia and met with the owner, who was asking $10,000 for seventy acres. Must offered $5,000. "That's right smart timber up there, Bobba honey," the farmer responded. (Appalachian men called their friends "honey" in those days, and some still do.) "I couldn't sell it for that." Bob offered $6,000, then $7,000, then $8,000, figuring he could borrow money, since he only had $7,000 to his name. The farmer held firm. "Well, thanks for talking to me," Bob said, shaking the farmer's hand and walking out of the house. Bob crossed the yard with his head down, trying to reconcile the fact that he'd lost his dream property. Then the door slammed back open. "Wait a minute, Bobba honey!" the farmer yelled. "I talked to my wife and she said we can go $8,000!"

Bob built a cabin using hand tools. It had no telephone, electricity, running water, or indoor toilet. He met Ginger in 1980 while visiting Boston, and the following year she moved in with him. After having two children, they built a bigger house with solar panels and plumbing. Bob used the G.I. Bill to pay his way through the West Virginia School of Osteopathic Medicine and worked in health care for about thirty years, including a decade operating a clinic in the nearby town of Hillsboro. Ginger became a librarian. When I later met their son, Andrew, he was converting an old Lobelia schoolhouse into a home for himself, and I asked why he hadn't branched off to find his own paradise, as his father had done. "I felt like he had already figured it out," Andrew said. "The environment is unspoiled and pristine. Humans haven't had much of a chance to ruin it. It's a nice legacy to continue."

And the hippies never really stopped coming. Through Yew

Mountain Center, I later met a rugged-looking character from New Jersey named John Leyzorek, who'd moved to Pocahontas in 1988 after going to the Princeton Public Library and studying maps for a place on the East Coast with no highways, cities, military bases, or nuclear plants. I hadn't realized so many people went to the library to figure out where to live. He purchased a six-hundred-acre parcel in central Pocahontas at $180 an acre. He later befriended a retired biologist from Maryland named Joel Rosenthal, who in the early 2000s purchased a 262-acre tract in southern Pocahontas where he started rehabilitating wild animals, from fawns to cubs. When the West Virginia Department of Natural Resources charged Rosenthal with illegal possession of wildlife, Leyzorek assisted him in a four-year legal battle that went all the way to the state supreme court, with Rosenthal ultimately winning the right to care for animals on his property, which he called Point of View Farm. The county had also been host to a well-known, 180-acre hippie commune called the Zendik Farm, which aimed to overthrow America's consumerist "Deathculture" and save the world from environmental catastrophe.

Had I walked into a dream? An elderly man was cohabitating with bears down the road from the world-famous clown doctor Patch Adams and just a few miles from a hippie enclave, all of them sharing a patch of Appalachia with world-renowned astronomers and secretive government operations. The area seemed tinged with magical realism, with an impossible menagerie of eccentrics congregating in the forest. How had so many random groups found their way to the same corner of West Virginia?

"All of these subcultures came here for a reason," said Dale, the former sheriff. To escape. To be left alone.

Not all the hippies settled permanently in Pocahontas. In 1984, Laurie Cameron found himself trying to sell a 267-acre parcel in

Mill Point that he'd initially purchased with friends to create a commune. After their plan fizzled, they put the land back on the market, and Cameron gave a property tour to an interested buyer who characterized himself as part of a church. During their walk-around, the man seemed to be sizing up the land's potential as a militia base, showing special interest in a ridge above the property. "He was interested in these heights and places that had wide fields of fire," said Cameron, himself an army veteran.

Just before selling the property, Cameron happened to read an article in the *Washington Post* about extremist groups in the United States and thought he recognized one of the people. Sure enough, it was William Pierce, the man with whom he'd spent seven hours walking around Mill Point.

Pierce had been on law enforcement's radar for several decades. In the '60s, he was a high-ranking member of the American Nazi Party led by George Lincoln Rockwell, a bombastic man with a penchant for "Heil Hitler!" salutes. After a former party member assassinated Rockwell in 1967, Pierce became a leader of Youth for Wallace, which supported the presidential bid of George Wallace, the segregationist governor of Alabama. In 1970, Pierce reconfigured Youth for Wallace as the National Youth Alliance, which became the National Alliance in 1974 to appeal to a wider audience. He started a monthly newspaper called *Attack!* that sold guns for "urban firefights," gave assassination tips, and provided bomb-making instructions. From 1975 to 1978, while in his early forties, Pierce wrote and serialized *The Turner Diaries* in *Attack!* Though poorly written even by Pierce's own judgment, the book quickly became a fundamental document for extremists, in part because of how it reframed white nationalism as a struggle against a tyrannical government. Inspired by the novel, some National Alliance members helped form a gang called the Order that, along with committing robberies and murdering perceived enemies, stole $3.6 million from an armored vehicle in California. A member of the Order later told FBI agents that he personally gave $50,000 in stolen funds to Pierce, who was suspected of having used the money to purchase the land in Pocahontas.

Cameron proceeded with the sale, but he also called the *Washington Post* to tip journalists off. Within weeks, according to Cameron, newspaper reporters and FBI agents had visited the area to keep tabs on one of the most dangerous hate leaders in America.

BACK ON THE FOUR-WHEELER, Pringle and I drove deeper into the National Alliance compound, coming to the organization's old meeting hall, surrounded by overgrowth. Inside, the smell of mildew was overpowering. No gatherings had been held here since at least 2010, Pringle said, though he claimed to have witnessed $80,000 in donations pour in during single meetings. When Pierce was alive, the National Alliance had hosted biannual summits with as many as 125 people.

The floor was lined with boxes of books, the kind you'd find at any bookstore. I spotted *Killing Kennedy,* by Bill O'Reilly, alongside *All Too Human,* by George Stephanopoulos. Atop one box was a battle shield, the kind actually used for combat, emblazoned with "an Aryan Nations symbol of some kind," according to Pringle. He said he was bringing in a construction worker who could rehab the building and turn the upstairs into a library and a workout gym. He claimed it would be the largest library in the county, even though most of the organization's books—an estimated twenty-seven thousand volumes—had already been relocated to Tennessee. Pringle said another building on the compound held boxes "stacked high to the ceiling" of *The Turner Diaries.* Considered by the FBI to be "the bible of the racist right," the book was credited with helping inspire a militiaman named Timothy McVeigh to bomb a federal building in Oklahoma City in 1995, killing 168 people and injuring another 680 in what remains the deadliest incident of domestic terrorism in U.S. history. "We're the preferred seller on Amazon," Pringle boasted of a book linked to at least forty terrorist attacks and hate crimes in the United States and overseas—the most recent of them the 2016 assassination of a British parliamentarian. He seemed to enjoy talking, and I didn't want to stop him, because I thought this might be my only chance to learn about the National Alliance from an insider. He was divulging far more than I'd expected to hear, and it was alarming.

As we revved along on the four-wheeler, Pringle pointed toward Pierce's former residence, a single-wide mobile home that I would have overlooked because it was surrounded by tall shrubs and trees. The floor was rotting out and the ceiling was caving in. This was where Pierce died of cancer in 2002.

In the nearby woods, Pringle had inoculated several logs with shiitake mushroom spores. He also showed me his homemade apiary. He said he wanted to make a sign that said "Our Bee Wants Free," a riff on the phrase "Arbeit Macht Frei" that had hung at the entrance to the Auschwitz concentration camp. Pringle thought it was funny because he found the whole idea of the Holocaust a joke—"Allied propaganda," he called it.

We rode another half mile uphill to the property's highest point, which had a panoramic view of surrounding farmland and mountains. There was a picnic table and a fire pit with charred beer cans, remnants of a gathering that Pringle had organized a few nights earlier. He said we stood on Pierce Point, where the late leader's ashes were scattered. Upon Pierce's death, the former Ku Klux Klan leader David Duke credited him with making "a tremendous contribution to our cause . . . After having read almost every word he wrote, I feel once more as though a family member was lost."

Standing atop Pierce Point, looking over the valley toward Droop Mountain, Pringle said he felt out of authority's reach, with the absence of cell service making it harder for the government to monitor his whereabouts. In a 1987 field report on the National Alliance, the FBI itself lamented that "surveillance opportunities are limited and infrequent due to the remote setting. Ground surveillance . . . is exceedingly difficult. Anyone from outside the area can be readily identified as a stranger by the local residents. Furthermore, the mere presence of strangers is information which is disseminated quickly throughout the area." The quiet was an asset to the neo-Nazis, a blanket for their ideology to hide under. Relocating to Pocahontas got Pierce away from law enforcement and hostile neighbors that badgered him back in Arlington, Virginia. In a place with a culture of minding one's own business, he could pursue his radical agenda in relative peace while preparing for a looming race war that he predicted.

"The thing about Dr. Pierce, he was a strategic thinker and he was thinking tactically," Pringle said. "He wanted a place he could defend. Even in the throes of violent revolution at lower altitudes, you couldn't get up here with an infantry division. And he wanted a place where if he had to grow his own groceries, he could. We're never going to run out of fuel. Look, it's all over the place," he said, waving to the trees. "Plus, there's food walking around everywhere. You kill a doe, a nice young yearling doe, it's like filet mignon."

I LATER LOOKED PRINGLE UP. He had once been on the Southern Poverty Law Center's list of "40 to Watch: Leaders of the Radical Right." (The list was alphabetical, so I didn't know where he ranked.) I'd had no idea he was so notorious, and I'd heard him strategizing how to bring more white nationalists into Pocahontas County and essentially teach them the religion of *The Turner Diaries,* which seemed to be experiencing a kind of sinister renaissance. Not a year earlier, the analyst J. M. Berger of the International Centre for Counter-Terrorism at the Hague had warned that the book was likely to gain new traction amid a "highly charged social climate" in which "mainstream politicians ratify white racial fear and white nationalist beliefs predicated on worries about terrorism and immigration." If the National Alliance really was staging a revival, I thought authorities would want to know, so I called up the former county sheriff David Jonese in Green Bank.

"I know a few guys who are sympathizers, they're not doing anything," Jonese told me, shrugging off the idea of the National Alliance posing any threat. Besides, he said, there was no law against being a neo-Nazi. "You can't legislate it, you can't change it, so you just learn to suck it up and live with it."

Jonese said his biggest concern in Pocahontas was drugs from a region-wide opioid epidemic. Only once during his eight-year tenure as sheriff had he responded to a problem at the National Alliance, when its chairman, Will Williams, was accused of assaulting a woman who worked there. Jonese had taken Williams into custody. "I talked to him for several hours while he was in prison," Jonese said. "He's actually a very cordial guy."

CHAPTER NINE

"A Low Roar"

CHUCK NIDAY HELD WHAT LOOKED like a futuristic stun gun from *Star Trek.* It allowed the Quiet Zone cops to find electric poles causing interference to the telescopes. The device picked up radio frequencies between 320 megahertz and 335 megahertz, which could point Niday toward staticky power lines. He next picked up a giant mallet.

"This is a troubleshooting technique for noisy power line hardware," he said as we stood outside the observatory. "You go up to the base of the electric pole and give it a good smack and look for the [radio] noise. If it gets worse, you know the noise is coming from somewhere on the pole. It's also incredibly dangerous because the noise might be caused by an insulator that's getting ready to break." Quiet policing had its hazards.

Niday put the mallet back inside the government-owned Dodge Ram 2500, an electromagnetic interference tracking truck known by the acronym EMITT. It reminded me of the wraith-hunting vehicle from *Ghostbusters,* and its aim was similar. Niday was effectively searching for ghosts: the invisible waves of electromagnetic radiation that are all around us, emanating from power lines and WiFi routers, flying through walls and zooming across the sky.

A bearded, bearlike man in his early sixties, wearing jeans and a ball cap, Niday was on the front line for maintaining Green Bank's radio-quiet environment. Hypothetically, you couldn't turn on a smartphone in town without him knowing. This had been his part-time role, in addition to other work at the observatory, since Wesley Sizemore's retirement in 2011. Niday was also a technician at Alleghany Mountain Radio, where his wife was the manager. (They cohosted a weekly jazz program that had about ten steady listeners—"pretty good for around here," Niday said.)

Seventeen antennas protruded from the truck's roof, a system for pinpointing local sources of radio frequency interference. A main antenna picked up signals from 25 megahertz to 4 gigahertz, while smaller antennas operated as a direction-finding array. "Through some method, which I believe involves witchcraft, it comes up with a direction for the signal we're looking for," Niday said. Every few weeks he went out on patrol to keep tabs on local noise levels.

It started to drizzle, so we hopped in the truck. Wiring snaked from the roof down to a stack of electronics and computer monitors in the cab. "Footloose" played on an AM/FM radio. Over a two-way communications radio, we heard the voices of bear hunters. Their dogs wore radio collars that had the potential to interfere with the telescopes, and these hunters appeared to be veering too close. But Niday had other concerns. He adjusted the dials on a computer to look for signals in the 2.4 gigahertz frequency: WiFi. He shifted into drive.

As we exited the observatory's parking lot, the monitor started bleeping angrily. Before we reached the main road, we had picked up thirteen wireless signals. Within a half mile, we found sixty-six signals. Niday's gadgetry was going berserk, revealing the radio pollution in Green Bank. We passed by Arbovale United Methodist Church, where Niday sang in the choir alongside Betty Mullenax, the elderly cashier at Trent's General Store across the street. We heard another bleep. Trent's had WiFi.

Instead of jumping out of the truck with a pair of handcuffs to bust WiFi offenders, Niday simply took note of the source of radio noise and kept driving. He didn't seem the least surprised. Within five miles of the telescopes, we counted more than two hundred WiFi signals, some coming from the homes of staff living on the observatory's own property—a blatant violation of the facility's regulations.

"It's not a radio silent zone," Niday said, rain beating down on the windshield. "We're just trying to keep everything down to a low roar."

I had always suspected Green Bank was not as quiet as media suggested, but Niday had just revealed a shocking level of noise. How could this be called the quietest place in America? Why had so many news stories portrayed this as a town without WiFi?

When I was next at Trent's, I asked manager Bobby Ervine if he'd ever considered installing WiFi.

He nodded.

"Do you have WiFi right now?" I asked.

Ervine nodded again, as if he were letting me in on a dangerous secret. This close to the telescopes, operating any wireless equipment likely caused interference and was arguably breaking state law.

"Does the observatory know you have WiFi?" I asked.

"You show me where it says in the law that I can't have it," he said, his tone turning defensive.

NOT LONG AFTER my patrol with Niday, CNN's medical journalist Sanjay Gupta visited Green Bank, driving into town for an episode of *Vital Signs*. "National Radio Quiet Zone," Gupta said to the camera, "that means there's no cell service, there's no WiFi, there's no radio. It's just really quiet." Soon after, the TV news personality Katie Couric visited for a National Geographic series. "Green Bank is a town where technology is almost completely banned," she said in a bright voice-over when the series aired, later opining, "People here seem happy to follow the law of the land."

Had Gupta and Couric so much as searched for a WiFi signal using their smartphones, they might have started to see a messier portrait of the Quiet Zone. Couric's 2017 visit included a stop into Trent's, which had installed WiFi a full year earlier. She and Gupta were each surrounded by supposedly banned tech. There *was* radio. There *was* WiFi. Even cell service had become negotiable within ten miles of the telescopes.

In 2015, Snowshoe Mountain Resort, which was about nine miles from the Green Bank Telescope as the crow flies, installed cell service through a specially designed system of 180 low-power antennas distributed around its ski village and slopes. The observatory helped oversee the installation, which reflected a complete reversal from its initial opposition to the mere idea of a ski resort. Back in 1974, the observatory's director had written a letter to Snowshoe's developer expressing concern about interference from the resort's communication system, ski-lift machinery, and automobile traffic. "Green Bank's high international reputation, and its value to the scientific community, are dependent in no small part on the 'radio quietness' of the site," the letter stated. "We are of course anxious that it remain quiet." Had something changed?

CNN and National Geographic were hardly alone in failing to scratch beneath the quiet facade. "This Town Lives Without Cellphones, Wi-Fi," the *Today* show reported in 2016. "Inside the U.S.'s 'National Radio Quiet Zone' Where There's No WiFi or Cellphone Service," a *Washington Post* headline read in 2018. I could somewhat understand how the media kept getting it wrong: strapped for time, unable to spend longer than a day or two in the area, a reporter might come with an assumption about what life was like and seek evidence to support that conclusion. But even public officials who should have known better played into the hype. "All people within a 20-mile radius of the facility cannot have any device that emits a noticeably high amount of electromagnetic radiation," Senator Joe Manchin would write in a 2018 op-ed. "This includes WiFi routers, cell phones, and even microwaves. Yet, these faithful West Virginians have sacrificed all of these luxuries for the advancement of science."

Green Bank resident Teresa Mullen, who taught at the high school, rolled her eyes at such language. She had a microwave. She had a smartphone. She had WiFi. "It's not like we're living some bohemian lifestyle," she told me. Such was hardly a secret. Even before joining Niday on his patrol, residents were admitting to me that they had all the supposedly banned electronics—and none said they'd been reprimanded. To be sure, *some* neighbors chose to not install WiFi, and a few even reached out to the observatory to seek approval before using any wireless equipment. But I found the opposite view far more prevalent.

A house across the street from the observatory had WiFi with the network name "Screw you NRAO," an unsubtle middle finger to the observatory's calls for quiet. Green Bank's health clinic had WiFi. Green Bank's senior center had WiFi. "We're not supposed to," said John Simmons, the county's director of senior programs and a former county commissioner, "but I think all that stuff about the noise levels is fabricated." Frontier Communications, the largest internet provider to the area, was egregious about installing WiFi routers, sparking frustrated phone calls from the observatory asking for the company's cooperation in protecting the Quiet Zone.

But it was useless. "At some point it becomes difficult to deny the local population the luxuries that other people have," as Sizemore put it. Trying to stop WiFi was like pissing in the wind.

Before retiring, Sizemore proposed getting more aggressive about wireless tech by reminding the community through newspaper ads and other public notices that the observatory had legal standing to request radio quiet. He considered the Quiet Zone the foundation upon which the telescopes were built; if the foundation eroded, the observatory would collapse. Management even considered prosecuting individuals who installed WiFi. But officials at the National Radio Astronomy Observatory headquarters nixed the idea. There was a reluctance to test the law or spend funds on litigation, as well as a desire to keep peace in the community. In the Quiet Zone's six-decade history, the observatory had never asked the county prosecutor to fine rule-breakers.

Sizemore's retirement led to a policy shift. Whereas he had hassled people to turn off WiFi and other electronics, his successors were instructed to merely take note of what they found and report back to the observatory's Interference Protection Group, co-led by business manager Michael Holstine. Sizemore had responded in real time to astronomers' complaints about radio noise and sometimes jumped into his car late at night or on weekends to chase down the source of interference. Now, the job mostly entailed passive monitoring.

"If they find a house that's particularly bad, then they'll bring that to my attention and I'll take care of it," Holstine told me. By that, he meant he might initiate a conversation with the offender and politely explain the observatory's need for radio quiet. "We lean on education to the community. We don't have the staff to do any further enforcement."

The lack of strict enforcement had led many people to believe WiFi was harmless. A mile away from the observatory, Rudy

Marrujo and his wife, Jan Cozart, had installed both WiFi *and* a mesh network to boost the signal into every room. Having moved to Green Bank from Silicon Valley, they had high expectations for tech connectivity. But they'd also hardwired their entire house with cable internet, on the chance they were ever told to turn off their WiFi. Once, around 2013, Marrujo spotted Niday pull up in his radio-tracking truck and point his gadgetry toward their house, but nobody ever said anything about their WiFi, which made them believe it was permitted. *If it was really a problem,* I was repeatedly told, *why doesn't the observatory do something about it?* "You can have WiFi in your home," said Oak Hall, owner of Red Oak Realty, which regularly sold homes around Green Bank and was a first point of contact for many newcomers. "You can have WiFi that reaches over to your neighbor's home. That's not something that's restricted, to my knowledge, in the Quiet Zone."

Another time at Trent's, I met a man named Daniel Solliday, a lay minister at New Hope Church of the Brethren. He was with his son, Mathias, who had already told me he didn't mind the lack of cell service or the rules against WiFi.

"You don't have WiFi at home, *right*?" I asked Daniel. I was learning to be skeptical.

"Right," Daniel said.

"Why are you holding out?"

There was an awkward silence.

"Well, actually, we do have WiFi," Daniel admitted sheepishly. He'd installed it in 2016.

I had to laugh—even a minister had lied to me about whether his home had WiFi. The Sollidays' internet was incredibly slow—far too slow for video streaming, for example—but they still wanted the convenience of WiFi.

"Why didn't you want to tell me you had WiFi?" I asked.

"It's a difficult thing because the observatory is important to our community," Daniel said. "Officially, you're not supposed to have it, right? But when the fast lane is going seventy-five miles per hour, that's the lane you have to be in."

If the Sollidays had been willing to lie, feeding into the quiet veneer around Green Bank, how much more tech might there really be?

"What the heck?" said a woman named Kathryn Stauffer, as if I myself had lied to her. "People come here thinking there's no WiFi because of the media . . . It's hurting people."

We were speaking at the observatory's Starlight Café, one of the "101 Unique Places to Dine in West Virginia," though we couldn't talk for long because Stauffer said she felt bothered by machinery in the kitchen. An electrosensitive, she had arrived in Green Bank in 2016, having traveled hundreds of miles from her home in Illinois, and the last thing she'd expected to find was WiFi, smartphones, and an occasional cell signal. She still believed Green Bank was quieter than most other places, though she had created a Facebook page called "Radio Quiet Zone, Green Bank" to warn people about the noisy reality.

Some electrosensitives were taking matters into their own hands, hunting down WiFi and demanding it be turned off. The sheriff's department had received dozens of 911 calls from one woman who complained that a neighbor's WiFi was killing her. In 2016, she walked into a county commission meeting wearing a full-body nuclear radiation suit, warning of the health dangers from WiFi and requesting the commission address it. She had a legal argument, too. The West Virginia Radio Astronomy Zoning Act made it illegal to operate any electrical equipment within ten miles of the observatory that interfered with the telescopes. The commission thanked the woman for her input and moved on to other issues. She left the county soon afterward.

As I dug into the issue, I found myself wading into a legal debate. Upon my initial arrival to Green Bank, Holstine had told me in no uncertain terms that the observatory could push back against any source of radio interference in town, be it WiFi or a smartphone, a microwave or a malfunctioning electric blanket. To me, the state law sounded clear enough: "It shall be illegal to operate or cause to be operated any electrical equipment within a two-mile radius of the reception equipment of any radio astronomy facility if such operation causes interference with reception by said radio astronomy facility of radio waves emanating from any nonterrestrial source." That act extended protections to a ten-mile radius, with a fifty-dollar daily fine for violators. Jay Lockman, the principal scientist and former site director, flatly said of WiFi: "It's illegal."

Then I spoke with Ken Kellermann, who'd been on staff with the National Radio Astronomy Observatory for a half century. He believed the state law applied only to "incidental, non-intended radiation," such as from a neon sign or electric fence. WiFi was an "intentional" transmitter and thus governed by the Federal Communications Commission and National Telecommunications and Information Administration. "State and local governments have no power to control intended radio transmissions," Kellermann said. "That includes WiFi."

While the federally designated National Radio Quiet Zone gave the observatory oversight over the installation of "permanent fixed location" transmitters, WiFi was not fixed—meaning the observatory had no legal power to restrict its use, agreed Harvey Liszt, spectrum manager for the NRAO and chair of the international Scientific Committee on Frequency Allocations for Radio Astronomy and Space Science. WiFi and other FCC-licensed transmitters, including microwave ovens, fell solely under the FCC's purview, he added. Moreover, Liszt and Kellermann questioned the very legality of West Virginia's law, as it infringed upon the FCC's jurisdiction.

I brought the debate to Anthony Beasley, director of the NRAO, which operated telescopes around the world and had a vested interest in maintaining the Quiet Zone. He agreed there was ambiguity to the laws. But he said the argument was somewhat removed from reality. Even if the laws could be interpreted in their strictest possible way, it still wouldn't make financial or logistical sense to hunt WiFi up and down the valley. A house with WiFi 10.1 miles from the observatory would cause the same interference as a house 9.9 miles away, so why crack down on one if you couldn't legally stop the other? For a cash-strapped observatory fighting to merely stay open, why hire lawyers to prosecute WiFi-users when that money could go toward scientific equipment, staff, and research?

"You've got to decide which hill you're going to die on," Beasley said. "Taking someone to court and potentially getting some kind of class action lawsuit going would be an incredible waste of time, in my opinion."

On that point, everyone agreed. It was impossible to stop the wireless revolution. Outside of the ten-mile radius, but still within Pocahontas County and the National Radio Quiet Zone, even the hippies were going wireless. Yew Mountain Center had WiFi, and its director, Erica Marks, carried an iPhone, allowing her to manage what was essentially a gig economy in the backwoods, juggling the center with parenting, teaching, and leading yoga classes. Lobelia was far enough from Green Bank for the technology to not interfere with the telescopes, but was it hypocritical to offer WiFi at a place that advertised itself as a way to get back to nature?

"It's been incredibly liberating to have a smartphone," Marks told me. "I could check out the trails and still be in contact with my family. It allowed me to stay out for longer periods of time knowing that things were fine."

There were downsides, however. Marks admitted that her two-year-old had developed an "opiate-like addiction" for watching *Peppa Pig* on the smartphone, and the toddler became enraged whenever the device was taken away. She later emailed me that, upon further observation, her daughter was equally furious upon losing access to a Sharpie. "It was somehow comforting."

DURING MY RIDE ALONG with Chuck Niday, we would have found even more signals had we driven a few miles farther toward his house. Even he had WiFi. "Technically" it wasn't permitted, Niday admitted, "but I know how to break the rules." By that, he meant he was fairly certain that his WiFi wouldn't interfere with the telescopes because he lived far enough away and behind a hill, though still within the sacred ten-mile radius.

In fact, many observatory staff had WiFi at their homes. An employee in staff housing was once discovered to have WiFi and was told to turn it off, per the terms of the rental agreement. The employee refused and was subsequently fired. WiFi was apparently more important than the job. Even Wesley Sizemore, the famed curmudgeon of quiet, had installed a wireless router before retiring from the observatory. He argued that his house—though only eight miles from the telescopes—would not cause radio interference, according to a propagation study that he'd conducted. And anyway, why should he sacrifice WiFi if the observatory wasn't going to crack down on a router across the street?

There was more. Since retiring, Sizemore had taken up a new profession as "senior RF designer" for the engineering firm TRC Solutions, with a role of "coordination of radio transmitters with the National Radio Quiet Zone," according to his LinkedIn page. He was helping wireless providers enter the area. He'd turned to the dark side.

On one of Sizemore's final patrols in 2011, he counted more than seventy WiFi hotspots within two miles of the observatory. Given that the 2010 census put Green Bank's total population at 143, it already begged the question: Who didn't have WiFi? In late 2017, Niday would detect 117 WiFi hotspots within two miles, a 70 percent increase over six years. The 2.4 megahertz frequency band had become so polluted that astronomers had lost access to that window into the radio universe. Rather than getting a clean reading of astronomical radio waves, a chart would show an imperceptible scribble of noise from the town's WiFi. "If E.T. calls on that frequency, we're not gonna hear him," Sizemore said.

The Quiet Zone was being breached. I felt that I'd stumbled into a pivotal place in the world and, perhaps, in the history of humanity: an area endangered not by climate change or gentrification but by the Fitbit on your wrist, the iPhone in your hand, the anti-collision sensor in your car, the human desire to have what everybody else has. Would Green Bank be able to preserve the quiet? And if it couldn't, what did that mean for my own quiet fight?

My objective had changed. Rather than finding a place where I might fit in, I was charting whether the quiet could survive for the sake of the astronomers and electrosensitives—and, rather unexpectedly, whether that same quiet was giving cover to white nationalism's reemergence.

PART TWO

QUIET DISCOVERY

> I am not at ease, nor am I quiet;
> I have no rest, for trouble comes.
>
> —JOB 3:26

CHAPTER TEN

"Local Nazi Diaspora"

QUIET IS RELATIVE. Once I adjusted to life in Green Bank, I could find myself distracted by a single vehicle passing outside my window, since my dead-end road was otherwise so untraveled. Back in New York City, I thought nothing of police cruisers, ambulances, and fire trucks blaring by, because it was all ambient noise. Quiet itself would have been disquieting.

I found it beneficial to reset my expectations by getting out of Green Bank. It was also necessary for maintaining a relationship with Jenna, who still lived and worked full time in New York City. And I had to scrounge up cash to fund my next visit to West Virginia. I was freelancing articles as well as substitute teaching at a Connecticut high school where students carried school-issued iPads; whenever I walked into the classroom, I'd find them streaming YouTube or playing video games. It was a helpful juxtaposition to Green Bank, where people were at least more discreet in their usage of devices on WiFi.

Jenna accepted my nomadic location-hopping and occasionally joined my forays back to the Quiet Zone, though she was growing skeptical of my aims. I'd pitched Green Bank as a kind of Walden, a disconnected place where we might "live deep and suck out all the marrow of life," in the words of Thoreau. Then I started coming back with stories of electro-allergies and illicit WiFi hotspots, secret government hideouts and neo-Nazi terror plots. The place was less and less Walden and more and more weird.

Back in Pocahontas County for another monthlong visit in the fall of 2017, I overheard a librarian in Hillsboro say she'd been seeing "the Nazis" a lot lately. They came to read the newspapers and magazines. More white supremacists had been spotted driving into the county, their identities revealed by their overtly racist bumper stickers. And there had been other signs of something bubbling below the surface. One person told me he'd found a neo-Nazi pamphlet shoved into a six-pack of beer. In 2015, "Nigger Lover" had been spray-painted on a restaurant in Hillsboro owned by a biracial couple. It all seemed to clue to a national trend, especially when hundreds of white nationalist protesters descended on Charlottesville, Virginia, for a Unite the Right rally that left one person dead in August 2017. An avowed neo-Nazi deliberately drove his car into a crowd of counterprotesters, killing a thirty-two-year-old woman and injuring twenty-eight others in what was deemed an act of domestic terrorism.

David Pringle, the chief of staff at the National Alliance compound in Pocahontas, tweeted an image of the car at the moment it struck several people. He included the line "meep meep . . . lefties make great bowling pins! Look at the bounce on a couple of those fools! #2010DodgeChallanger #TakeTheDodgeChallange." (He apparently forgot to spell-check "challange.") In making a joke out of violence, just as he'd joked about the Holocaust when I'd met him earlier that summer, Pringle was being deadly serious about his embrace of a racist agenda. The Charlottesville clash came amid a 57 percent surge in anti-Semitic incidents in the United States that year, a 17 percent jump in hate crimes, and a doubling of murders attributed to white supremacists to eighteen.

I scanned photos from Charlottesville, looking for Pringle among the faces, unable to shake the idea that the roots of America's modern white supremacist movement traced back to a mountainside hideaway in the Quiet Zone.

THE FIRST TIME I'd visited the National Alliance, the entrance gate was shut, a chain wrapped around it. The second time, the gate was wide open. I drove through and parked outside the tan office building adorned with the life rune symbol. Pringle stood outside wearing faded jeans, hiking boots, a grungy T-shirt, and a gray cap. His leg was in a brace.

He never made it to the Unite the Right rally. A week before the event, he'd been driving a four-wheeler when his foot slipped off a peg and got pulled under the wheel. He broke his tibia-fibula and badly bruised his hip from falling on a .40-caliber Smith & Wesson pistol tucked in his pants. He was rushed to a hospital in Roanoke, Virginia, over two hours away. By the time of the Charlottesville rally, Pringle was back at the compound on a couch, his broken leg propped up as he watched a livestream of the event over patchy internet with his wife.

"It's like I didn't get to go to the prom," Pringle said of missing Charlottesville. "I was supposed to be rolling in with Matt Heimbach and the Traditionalist Worker Party." Heimback was a rising leader in the alt-right at the time, though he later renounced white nationalism. "A big conversation we had amongst ourselves here beforehand was, are we going to open carry? And if we were, what was our threshold going to be before we did something with it?"

Other National Alliance members had attended Charlottesville, including a man named Albert Hess, who went by Jay. As I spoke with Pringle, Hess came outside, shirtless. Stout and round-faced, he wore blue Corona-branded sweatpants and sandals. He'd recently moved to West Virginia from Florida with his cats. Decamping to Appalachia was a way to clear his head after his wife's death, he said. He'd brought his drum kit as well as a mounted butterfly collection that he proudly showed me.

"It was crazy," Hess said of Charlottesville. "The police basically let everybody have at it."

For the National Alliance, Charlottesville was successful in at least one regard: sales of *The Turner Diaries* spiked afterward. Pringle estimated a "couple hundred" copies sold in the following weeks, compared with the more normal dozen a month. He claimed that new members had also joined the organization, paying the twenty-five-dollar application fee and twenty-dollar monthly dues. VICE News had inquired about visiting the compound, though Pringle turned the reporter down. He would only allow photography and video once he had everything "extra, extra, extra tidy."

I saw his point. The place looked as dilapidated as ever, the Appalachian forest slowly reclaiming the land, even as Pringle tried to convince me that he was making steady progress rehabbing the property and rebuilding community support. He'd recently recognized a man at the grocery store wearing a swastika belt buckle. "I call him the local Nazi diaspora," Pringle said. The National Alliance's former chief financial officer, Robert DeMarais, still lived close by, as did the former books division president, David Sims, among around a dozen people in the area once associated with the organization, some of them still active in the Far Right. Another former Alliance member living in the Hillsboro area, Alan Balogh, would later make news for helping start a new white nationalist political organization called the National Justice Party. He and his son, Warren Balogh, were founding council members alongside the alt-right personality Mike Peinovich. Pringle said he was still hosting monthly barbecues for locals.

He was also keeping busy building a rifle, a "fun" project to try to replicate one of the longest-ever sniper kills, a distance of about 1.5 miles. He'd been working on the gun since I first saw him in June, and in the meantime a new record had been set for a sniper kill from just over two miles. A professional gunsmith with a degree from the Colorado School of Trades, Pringle said the parts totaled about $14,000, with the cost split with two other friends. He was recording footage of himself assembling the rifle with music from a heavy metal band called Lamb of God. "They're total lefties but it's got this great guitar," he said, breaking into an air guitar solo.

Pringle made gun-building sound like a harmless hobby, though in the same breath he described his profession as "sharpening the Grim Reaper's blade" and said he'd once given advice to a man named Jason McGhee on what kind of gun to purchase for the "specs" he was seeking—not that he knew McGhee would go on a murderous rampage with the weapon, he added. A former National Alliance employee, McGhee in 2006 shot and stabbed four people at a youth hangout in Georgia, killing them and injuring three more people. McGhee was one of three people charged in the attack and sentenced to life in prison.

Pringle invited me inside to see the rifle, retrieving it from a metal cabinet. A gigantic scope mounted to the barrel could aim to within an inch of accuracy from a mile away, he claimed. This was no hunting gun. He handed it to me. Cold to the touch, the thing weighed at least twenty pounds. In the moment, I tried not to think about the weirdness of the situation: me with a neo-Nazi, handling his sniper rifle, leaving my fingerprints everywhere before it was used for God knows what.

Since we were inside, Pringle offered to show me William Pierce's former study. He led me through a couple of doors. "We referred to it as the Sanctuary," he said, pointing to where there once sat a large desk, a shelf of research books, and a giant leather-bound dictionary. The room was now mostly empty, scattered with old metal furniture, scrap paper, and some decades-old National Alliance trucker hats that Pringle had heard were created by former member Bob Matthews, who had led the Order, the deadly white nationalist gang. About two decades earlier, *60 Minutes* correspondent Mike Wallace had interviewed Pierce in this very room and inspected his first edition copy of *Mein Kampf,* the autobiography of Adolf Hitler. By then, Pierce's novel *The Turner Diaries* had sold almost two hundred thousand copies and become, according to Wallace, "the bible of Far Right militiamen like Timothy McVeigh, who is now awaiting trial for the Oklahoma City bombing."

McVeigh had been obsessed with *The Turner Diaries,* gifting it to friends, selling it at gun shows, and always carrying a well-thumbed copy. Pages from the book were found in his getaway car from Oklahoma City. A copy that he'd given his cousin was the government's exhibit no. 1 in *United States of America v. Timothy James McVeigh.* He'd also called a National Alliance hotline seven times the day before the bombing. When McVeigh was later put to death, Pringle sent a mass email in defense of the mass murderer, saying McVeigh "should have a monument erected in his honor." Pringle added that he personally felt no "sympathy for the families of the 168, not the children, not the secretaries, and definitely not the federal pigs." He seemed to feel some shared history with McVeigh, as they'd both been first exposed to organized white nationalism while serving in the army in the late '80s.

While Pierce repeatedly argued to the press that *The Turner Diaries* could not be seen as a blueprint for McVeigh, he still credited the young army veteran with drawing new recruits to his organization. The National Alliance, which had three hundred associates when Pringle joined in 1992, reportedly amassed twenty-five hundred dues-paying members in the years after the bombing, although that may be an overestimate. Former deputy membership coordinator Billy Roper told me membership peaked at 1,274 people in 2001.

By all accounts, at the turn of the century, the National Alliance was one of the world's most influential white power organizations. Pierce commanded forty-three units in twenty-six states, plus another five units in Canada, according to Leonard Zeskind's *Blood and Politics,* a history of white nationalism in America. The National Alliance had expanded into a hate media empire, with twenty staff and $1 million in annual sales. Along with publishing a monthly magazine and broadcasting a weekly radio show called *American Dissent Voices* to one hundred thousand listeners, Pierce operated a music label called Resistance Records that sold albums by bands with names such as Angry Aryans and White Wash. He also launched *Resistance* magazine—called "the *Rolling Stone* of the hate music world"—and branched into video games. In one, called *Ethnic Cleansing,* a city-roving protagonist must kill Blacks, Latinos, and Jews, with the game's final challenge being to kill the Israeli prime minister. National Alliance ads were appearing on billboards and buses around the country.

Pringle, who led units in Albuquerque and later in Anchorage, first met Pierce in 2000 when he attended a leadership conference at the compound. During one event, Pierce cut off Pringle's ponytail as part of a fund-raising drive.

"I still have it," Pringle said of his severed lock of hair. "I'll show it to you if you want."

"Sounds gross," I said.

"It's kind of a weird thing," he admitted.

We returned to the front of the building, where Pringle kept a small office. The phone rang. He limped behind his desk to pick up the corded handset. It was his dad—a reminder to me that the guy was still human after all. I glanced around. A bookshelf held copies of Pierce's personal Bible, of *Mein Kampf,* and of *My Awakening,* by David Duke, the former Louisiana state congressman and KKK leader. There was a beer koozie with the words "David Duke U.S. Senate." Duke was a personal friend, Pringle explained when he got off the phone. Pringle seemed to have a lot of infamous friends.

Another acquaintance was Dylann Roof, the young man awaiting execution for killing nine congregants at a historic Black church in South Carolina in 2015, including the pastor and a state senator. From prison, Roof wrote a letter to the National Alliance asking for reading material for a "political prisoner." Roof said he'd already read *The Turner Diaries,* so Pringle mailed him some issues of *National Vanguard* and was now planning to send him a copy of *The Lightning and the Sun,* by Savitri Devi Mukherji, a French-born convert to Hinduism and Nazism who believed that Hitler was an incarnation of the god Vishnu. The book was published by the National Alliance. Pringle was including a self-addressed stamped envelope so Roof could write back. "I'm a nice guy, you know?" Pringle laughed. "And he's a limited-time pen pal." (Roof was also in contact with Roper, who had started another white nationalist organization in Arkansas; Roof asked for details about what it was like to work with Pierce, underscoring the young man's fascination with the hate leader.)

Pringle quickly added that he didn't approve of Roof's actions. As a roundabout way of explaining himself, he cited the book *Essays of a Klansman,* by Louis Beam, a strategist of the white nationalist movement who created a point system for how to become a so-called Aryan Warrior. It entailed killing either one thousand Blacks or other nonwhite persons, twelve journalists, five members of Congress, or one U.S. president, among a number of murderous combinations. Pringle's conclusion was that "clearly, running around and shooting Blacks is not the way to power."

"Moral justification and tactical justification are two different things," Pringle said.

I asked, "So what Dylann Roof did wasn't bad because he killed Blacks but because it wasn't tactical?"

"Right, he gave in to indiscipline . . . Is it at a point where Roof needs to get his Glock 21 Slim Frame and go into the room and kill everybody? No. Will we be there eventually? That's not for us to decide. The Blacks will decide that." He spoke over a din of crickets.

AFTER PIERCE'S DEATH IN 2002, Pringle became the membership coordinator for the entire organization, considered the second-in-command. He quit in 2004 amid interorganizational squabbling and over the next decade started a website called White Wire and migrated through various contingencies in the "alt-right milieu," as he described them. After attending the 2016 funeral of a Utah rancher named LaVoy Finicum, who died while participating in Ammon Bundy's armed occupation of a national wildlife refuge in Oregon, Pringle felt motivated to reengage with the National Alliance, which had gone through a leadership change.

His return to Pocahontas in May 2016 coincided with the political rise of Donald Trump, which didn't seem coincidental. Trump's presidency was endorsed by white nationalists, including David Duke. After the Charlottesville rally, Trump claimed there were "very fine people on both sides," essentially giving a pat on the back to the white supremacists. I assumed that Pringle was optimistic about the National Alliance's future under the new president.

Instead, Pringle said Trump was giving disgruntled white people less motivation to join the movement. "His election makes people think that the solution is simply getting the right people in office, and it's not," Pringle said. Which was why he was working to recruit new members by distributing outreach pamphlets around the country. He called it "cadre development." He gave me a copy of a flyer, which had an image of an angelic white woman with the words "Love Your Race." Pringle had such canvassing down to a science.

"Don't take this personally, but most journalists are really lazy," he said. "They like things to land in their lap. I'll put out some flyers and then call in a complaint." The tactic ginned up free media coverage. He said he'd done it for years and at all kinds of events, including the 2002 Winter Olympics in Salt Lake City.

More ambitiously, Pringle said he was planning an alt-right festival in Pocahontas County for April 20, 2018, on Hitler's birthday. He planned to invite prominent white supremacists such as Duke who could pull in a crowd and show the potential economic benefits of a thriving National Alliance. Or at least that was his thinking.

Before I left, he gave me copies of *The Turner Diaries* and another Pierce novel, *Hunter,* about a white racist who murders interracial couples and spurs copycat acts of violence. Pierce had dedicated *Hunter*—which has sold more than a half million copies since 1989—to an American Nazi Party member named Joseph Paul Franklin, who had roved the country killing Jews, Blacks, and people in interracial relationships. Franklin also attempted to murder the magazine publisher Larry Flynt for displaying interracial sex in *Hustler,* leaving him paralyzed from the waist down. Franklin was executed in 2013 in Missouri after being convicted of eight murders between 1977 and 1980, but he claimed responsibility for at least a dozen more killings—including two in Pocahontas, not far from the National Alliance compound.

Pringle mentioned that he'd looked up my name and read my blog. A lump formed in my throat as I realized I'd posted photos online of hikes with Jenna, who is Korean. Was Pringle hinting that he knew I was in an interracial relationship? In *The Turner Diaries,* on something called the Day of the Rope, tens of thousands of people are hanged from lampposts, power poles, and trees, each with a placard around the neck that says "I defiled my race." In Pierce's nightmarish vision, Jenna and I would both be strung up. In Franklin's real rampage of terror, he simply used a rifle to achieve the same end. What kind of danger had I exposed myself and Jenna to? She never wanted me to visit the National Alliance. Now we both had fallen under its gaze. The moment made me recognize that I know nothing about the insidious threats that Blacks and other minorities endure on a daily basis.

It struck me that Pringle was assembling a gun that Franklin could only have dreamed of deploying. How many more people might Franklin have killed with the right weapon? And could the chatty, limping man before me carry out the same type of violence?

CHAPTER ELEVEN

"Command the Evil to Be Gone"

I WAS TOLD TO NOT WORRY about the National Alliance. *They're harmless,* people said. *They're mostly gone. Don't focus on them.* I didn't want to give oxygen to a hate group. But I also felt I couldn't ignore the flame of racism that had burned in Pocahontas County for so many years. If William Pierce had thrived in the Quiet Zone, what was to stop his organization from mounting a comeback?

Not unlike David Pringle, Pierce had presented himself as likable enough, well-read and occasionally eloquent, which could mask the vileness of his beliefs. Tall and lanky at six feet four inches, often dressed in jeans and a cardigan, with a trim haircut and thick spectacles, Pierce considered himself an academic and preferred to be called "Dr. Pierce." Originally from Georgia, but raised in Alabama and Texas, he held a doctorate in physics from the University of Colorado and was formerly a tenured professor at Oregon State University. He dismissed the Nazi salutes and Klan costumery of his predecessors as clownish and unserious, and he tried to remake the white supremacist movement into a more palatable, mainstream organization through the National Alliance (itself a bland and unassuming name). His unthreatening manner bought him some breathing space in Pocahontas.

When Pierce first arrived, there was some local pushback. I found a 1985 letter to the *Pocahontas Times* written by a church congregation near Green Bank that read: "Racism, bigotry, violence and methods aimed at intimidation are contrary to the values of Christianity and to the values of most of the residents of West Virginia and Pocahontas County . . . We ask the leaders of our state and county government to take whatever actions that are constitutionally right and proper to encourage those in the Cosmotheist Community and the Neo-Nazi National Alliance to leave our county and state."

Alarmed that dangerous people were coming into the county, Sheriff Jerry Dale spearheaded state legislation to curb paramilitary training in West Virginia, which he saw as a way "to prevent the Mill Point group from starting a terrorist training camp here in Pocahontas County," as he wrote in 1988 in the *Pocahontas Times.*

Over time, however, Pierce ingratiated himself in the community. He shopped locally. He dined out. He exchanged pleasantries. He dated a hippie from Lobelia. His associates also dated locally. Will Williams, who lived on the compound in the '80s and '90s, went out with a woman from Green Bank who showed him around the observatory. In turn, he brought her to a gathering at the compound, where Pierce often walked around with a Siamese cat perched on his shoulder.

"He was just a gentleman," recalled Joseph Smith, president of the county's historical society and a former mayor of Marlinton, who worked at a meat shop that Pierce frequented. "If you didn't know who he was and what he was connected with, you wouldn't think any more about him." Sam Felton, the current mayor of Marlinton, recalled greeting Pierce when they passed each other at French's Diner in town. "A lot of people that had no knowledge [of the National Alliance] would say 'Oh they're good people, they're just like us,'" Felton said. Eugene Simmons, the longtime county prosecutor, bumped into Pierce at the Hillsboro Post Office, which saw a huge increase in incoming and outgoing mail because of the National Alliance. "Pierce was a pretty sharp fella," said Simmons. "He stayed in his own area and nobody bothered him." Pierce wanted it that way. Most people "leave us alone as long as we don't bother them," he wrote in a 1988 news bulletin to members. To many locals, Pierce was simply minding his own business up on the hill.

"When Dr. Pierce was alive and running the operation, it was pretty quiet," said Jeff Barlow, who became county sheriff in 2017 and was a state police trooper based in Marlinton for twenty years before that. "If someone from the organization got in trouble, they came down, paid the fines, and Pierce sent them out of here. He didn't want trouble."

It was all a facade. In reality, Pierce was essentially running a criminal organization out of Pocahontas County. "The fact is, it was a gangster operation which produced an enormous number of murderers, bank robbers, and people who engaged in similar activities," said Mark Potok, the former editor-in-chief of the Southern Poverty Law Center's quarterly *Intelligence Report*. "This was essentially a nexus of criminal activity along with a healthy dose of really vile fascist ideology."

To wit, in 1996, white supremacists robbed $20,000 from a Connecticut bank and then hand-delivered a portion of the money to Pierce at the compound. Hardly a "gentleman," Pierce was a misogynist who consumed pornography "in a huge way" and spent the remainder of his time calling for genocide against the many people he hated, according to Potok. In 1990, Pierce was arrested for beating up his bookkeeper, which fit into a pattern of abuse against his own family. His son Kelvin Pierce has written an autobiography called *Sins of My Father* that reveals in brutal detail the merciless beatings his dad exacted with a belt, an electric razor cord, and a two-by-four. Prone to fits of volcanic rage, Pierce killed the family's two cats, snapping the neck of one and slamming the other against a wall. He hid bomb-making materials and weapons in the family's Virginia home, only narrowly escaping detection from FBI agents. He taught Kelvin how to make a pipe bomb and attempted to indoctrinate him into denying the Holocaust and admiring Hitler. When Kelvin was in high school, his father walked out on the family for West Virginia.

In Pocahontas, Pierce established a community of like-minded hate. To be invited to his compound was to enter into the company of an elite cadre of the white supremacist movement. "We did have a sense of community there that, really before the internet and social media took off, was lacking for a lot of white nationalists in terms of having the opportunity to socialize and to have the echo chambers where their political views were reinforced and further radicalized," according to Billy Roper, who worked at the compound from 2000 to 2002 and was briefly considered a potential successor to Pierce.

Far Right leaders from around the world journeyed into Appalachia to visit the National Alliance. Members of Germany's fascist National Democratic Party made the pilgrimage to Pocahontas. The neo-Nazi musician Hendrik Möbus, who served time for murder, hid at the compound for several weeks in 2000 while fleeing an international arrest warrant for violating the terms of his parole. Another visitor was the British expatriate Mark Cotterill, leader of a group called American Friends of the British National Party, which was funneling money to white supremacists in Britain. "The whole network was tightly connected to National Alliance members and to Pierce," according to Heidi Beirich, who for many years led the Southern Poverty Law Center's Intelligence Project and did undercover monitoring of Cotterill's organization.

Don Black, founder of the white supremacist radio show and web forum *Stormfront*, recalled a fairly cordial welcome when he visited Pocahontas in February 2002, several months before Pierce died. He made the trip with a British politician named Nick Griffin after they attended an American Renaissance conference in Washington, D.C. Griffin was then leader of the British National Party; he would later become a member of the European Parliament. Black and Griffin stayed at Graham's Motel, where Pierce met them. "The townsfolk were all friendly," Black told me.

Graham's was the go-to overnight accommodation for white nationalists, as it was the closest lodging to the National Alliance, about five miles away. The organization fully booked its eight rooms two weekends a year for conferences. Local law enforcement, aware of the activities, regularly monitored the motel and took down license plate numbers.

"When they were having a conference and I knew the conference was coming up, I didn't take reservations from anyone else," said Jaynell Graham, who ran Graham's Motel. "I did not have other people stay there that might be offended by anything they might overhear, because old motels have thin walls." Graham suggested that Timothy McVeigh may have also stayed at her motel, as her mother believed she recognized him on television after the bombing; Pierce himself repeatedly said he'd never met McVeigh.

Pringle told me he used to stay at Graham's "all the time." ("They had an okay breakfast," he added.) He also booked the entire Marlinton Motor Inn for National Alliance conferences. A popular hangout was a dive bar in Marlinton where a skinhead's girlfriend worked. When the National Alliance rented the bar, the owner put out a letter board sign that read "Welcome Nazis." (A photo of the sign was published in *Resistance* magazine.) Roper recalled another bar near the compound being a contact point for locals who wanted to learn about the National Alliance. Through the bar, Roper sold KKK patches, "hatecore" CDs, and autographed copies of *The Turner Diaries*. "The local redneck guys got along reasonably well with the skinheads," Roper said. "There was a lot of nudge-nudge, wink-wink implicit racial attitude among the locals."

I'd heard firsthand about discrimination in the area. Blacks heard whispers behind their backs and felt outright hostility. Puerto Rican migrant workers heard people mutter "fucking Mexican" at them. A fire chief, knowing full well that I was recording our interview, referred to the Obamas as "monkeys." At first I'd thought I was just encountering the pernicious racism found almost anywhere. But things apparently went deeper. Roper's impression probably also reflected how most Pocahontas residents never had to worry about being hate targets. The community today is 97 percent white, largely made up of Christian conservatives who overwhelmingly voted for Donald Trump. It could be easier for them to be more dismissive of the National Alliance than, say, a Jew, a Black, or an outspoken liberal.

To some people, the white supremacists were just another clientele. "It wasn't a big deal because they kept to themselves," said Graham. Their ideology didn't really matter.

She pointed to the former Ku Klux Klan member Robert C. Byrd, the state's longtime senator. "That would not fly now," Graham said of Byrd's controversial background. (Byrd called his early involvement in the KKK his "greatest mistake.")

And unlike skiers who stayed at Graham's Motel during trips to Snowshoe, the National Alliance folks were always tidy. "They never even left their fingerprints," she said.

Listening to Graham talk, it was easy to think of Pierce and his ilk as polite and harmless, which only underscored their subtle threat. "Overt displays of hate could be countered, prevented, or ignored," as the journalist Seyward Darby notes in the book *Sisters in Hate,* which profiled women in the white power movement. It's harder to fight "white nationalism's quotidian allure." I later learned that a National Alliance leader had violently assaulted his girlfriend in a room at Graham's Motel.

As the National Alliance grew in infamy, Pocahontas acquired a reputation. "People would say, 'You're the ones that have the Nazis!'" recalled Allen Johnson, director of the county's public libraries from 2001 to 2012. An activist at heart, Johnson saw the libraries as an avenue for engaging and strengthening the community, "which meant getting outside the walls and not just hoping somebody comes in to check out a book," he told me. It also meant using the library as a platform to publicly confront the National Alliance.

To be sure, Pocahontas County had not invited the white supremacists, and Pierce would likely have thrived in any remote location despite any amount of denouncing from the local population. But Johnson felt compelled to at least speak up.

"The community's attitude was to live and let live," Johnson said. "And I'm not that kind of guy."

JOHNSON GREETED ME on his porch with a big smile, blood under his fingernails and animal guts on his jeans. He'd just slaughtered fourteen of his wife's rabbits for their meat. Six-foot-one with a white beard, Johnson wore a yellow T-shirt and rubber Crocs with socks. He led me across the muddy yard to a tree where, after breaking each bunny's spine with a twist of the neck, he pinned their hind legs to two nails, beheaded them, then skinned and gutted them. Blood pooled at the bottom of the tree.

"You actually have some meat on your pants there, Allen," his wife, Debora, said. "It's probably a piece of fat."

He flicked it off and showed me around the yard, which was dominated by a pond stocked with largemouth bass, bluegill, and channel catfish; muskrat and snapping turtles also found their way in. Surrounding the pond were gardens of vegetables, flowers, and berries, hedged in by national forest. Johnson and his four sons had fished in, swam in, and skated on the pond, which had an island with a small wooden duck house. A few years earlier, Johnson had decided he didn't want ducks pooping and shedding feathers around his yard anymore, so he gave them away. Then geese moved onto the island and started eating all his vegetables, so Johnson and his golden retriever, Maggie, chased them away. It was a persistence he'd also deployed against the neo-Nazis.

"I've still got some rabbits to cut up," Johnson said, as I followed him inside. The kitchen table held a bucket of still-warm rabbit meat alongside a jackknife, a butcher knife, and a wood block. He demonstrated how he severed the hind legs, then the forelegs, then the neck and tailbone.

The slicing and dicing continued into the evening as Johnson told me how he and Debora—both from Indiana—had moved to Pocahontas in 1976, when Debora took a teaching job in the county. Soon they attended New Hope Church of the Brethren, led by David Rittenhouse. They found the service so welcoming that they purchased nine acres next door.

Not a decade later, the neo-Nazis also moved into the county, and the Johnsons watched as the National Alliance grew into the largest and most influential white supremacist organization in the country. At a local gas station, you might bump into young men wearing combat boots and swastika armbands, part of the more overt neo-Nazism associated with Resistance Records. Getting change at the store, a dollar bill might be stamped with the organization's website address. Coming out of a restaurant, a National

Alliance brochure or CD might be tucked under the car windshield wiper. Homes displayed the tricolor national flag of the Third Reich. Parents would go on playdates and discover other parents displayed swastikas in the house. Kids dared each other to call the National Alliance during sleepovers, a way of freaking themselves out. Some local youths were recruited into the organization.

The neo-Nazis also became a presence at the county's public libraries. Johnson recalled how Roper—dubbed "the uncensored voice of violent neo-Nazism" by the Southern Poverty Law Center—used the public computers in Marlinton's library in 2002 to create a website called White Revolution. Johnson felt he couldn't stop Roper from using a public facility, but he did tell Sheriff Jerry Dale where to find Roper's internet browsing history.

For more than a decade, Dale had worked with the FBI and other national agencies to keep tabs on Pierce, developing informants within the National Alliance and conducting periodic stakeouts. Once, when Pierce met with several dozen associates at a restaurant in Marlinton, Dale set up a camera in a nearby window to get photos of attendees and their vehicles. In a 2000 article about Pierce and his acquisition of Resistance Records, *Rolling Stone* described Dale as "Pierce's official hometown nemesis." Pierce, in his newsletter, accused the sheriff of "grandstanding for the Jewish media."

Roper had been trying to build links with other hate groups, including the Creativity Movement, whose founder was friends with Pierce. Soon the Creativity Movement also gained a foothold into Pocahontas County, when a man named Craig Cobb arrived around 2003, moving down the road from the Johnsons and opening a shop called Gray's Store, Aryan Autographs and 14 Words, L.L.C. It was a reference to a fourteen-word catchphrase of the radical Right: "We must secure the existence of our people and a future for White children." The shop displayed Cobb's hate literature and propaganda, which he was known to distribute locally along with "hatecore" music. (Cobb would later attempt to create an Aryan settlement in North Dakota, as documented in the 2015 film *Welcome to Leith*.)

A gangly, beady-eyed man with long blond hair, Cobb became a regular presence at the Marlinton and Green Bank libraries. He often carried a video camera, recording his interactions with Johnson and threatening to sue anyone who tried to curb his First Amendment rights to free expression. Parents asked the schools to stop bringing students to the libraries. "They were worried their kids would get proselytized into Nazism," Johnson said.

Frustrated with the situation, Johnson launched an educational campaign about the National Alliance. With input from the civil rights division of the state attorney general's office, he created a library web page that called out the National Alliance as a racist organization and tracked its members' actions. He hosted informational sessions to raise awareness about the group. He stocked the library with nearly fifty books about the white nationalist movement and organized them all into the Pearl S. Buck E Pluribus Unum Collection, named after the first American woman to win the Nobel Prize in Literature and notable as an advocate for minority rights and mixed-race adoption. Buck happened to have been born about five miles from the National Alliance headquarters.

In January 2003, Johnson organized a "unity march" through Marlinton with several dozen locals carrying a sign that read "Pocahontas County United Against Hate." In a bulletin that Johnson created for the march, he wrote: "In the past two years many dozens of National Alliance supporters have flowed in and out of the county to work with the self-termed 'hate core' music label, Resistance Records . . . The rally seeks to begin a coordinated process to strengthen the resolve of all citizens of good will to treat all people with kindness, respect, and courtesy."

In large part because of Johnson's efforts, Pocahontas in 2003 received one of six National Medals for Museum and Library Service from the U.S. Institute of Museum and Library Services in Washington, D.C. First Lady Laura Bush personally presented the award to Johnson.

While his work was commendable, what most undermined the National Alliance was the National Alliance itself. Following Pierce's death, the organization imploded amid a power dispute, mismanagement, and infighting. "As an activist political organization, the National Alliance fell into dysfunction almost immediately after Pierce's 2002 death," according to Potok, the hate groups analyst. Graham's Motel, which used to be fully booked two weekends a year with well-dressed white men attending the National Alliance's biannual conferences, shut down in 2005. The same year, Jaynell Graham joined the *Pocahontas Times* as a copy editor, later becoming the editor in chief. She sold her motel's mattresses and beds to the Old Clark Inn in Marlinton, where Jenna and I once stayed—meaning we'd likely slept in the same bed as a card-carrying neo-Nazi. (The National Alliance still issues physical membership cards.) That we'd gone to Green Bank in search of a peaceful place, only to find ourselves touched by the area's darker history, underscored the National Alliance's entwinement with the community.

"I believe the collapse is because the people rose against it," Debora Johnson said of the National Alliance. We sat at her kitchen table, her husband still dismembering rabbits. "When people drove by the entrance, they would command the evil to be gone."

"It's like Whac-A-Mole," Allen Johnson said. "They pop up somewhere else."

He now had other fights on his hands. An environmental activist, he had helped form an organization called Christians for the Mountains that protested a form of coal strip-mining called "mountaintop removal." His hero was Larry Gibson, an anti-mining activist from West Virginia. Johnson himself had been arrested for demonstrating in front of the White House and trespassing during an anti-mining rally near Beckley, West Virginia. His latest campaign was against a major natural gas project called the Atlantic Coast Pipeline, which was to pass within a half mile of his house.

Whack! Johnson continued chopping through rabbit bones. Maggie the dog waltzed into the kitchen chewing on a discarded rabbit head.

"Oh, Maggie!" Johnson cried.

He tried to grab a rabbit ear flopping from Maggie's jaws. She snarled. He dragged her outside by the collar. I felt nauseated from the sight of it all.

It was dark when I got back in my car. My headlights illuminated the rear bumper of Johnson's truck, which had two stickers. One read "I love mountains." The other said "I love my wife." Debora had added the second sticker.

Against a bright sky of stars, the full moon cast shadows into the dark forest. Fog floated over the hayfields. Deer bobbed against my headlights. A dead owl lay in the road. It was hauntingly beautiful.

CHAPTER TWELVE

"Murder by WiFi"

TWO-THIRDS OF POCAHONTAS COUNTY residents had library cards, an indication of how public facilities took on outsized importance in a remote community with minimal social gathering spots. While Allen Johnson was happy as library director to see everyone use the county's resources, some idiosyncratic characters tested his patience.

As the white supremacists retreated from Pocahontas County, a new group emerged. Diane Schou and the electrosensitives became a daily fixture at the Green Bank Public Library, taking advantage of the computers and free internet—so long as the overhead lights were turned off, because of their complaints about feeling pain from the lightbulbs. Computers and lights both give off electromagnetic radiation, but Schou and others argued that it was better to reduce their exposure however possible. They could apparently get by in darkness, but not without internet.

Schou once gave Johnson a booklet about electromagnetic hypersensitivity to add to the library—she donated a similar book to the National Institutes of Health library outside Washington, D.C.—but he declined to add the title to his county's collection. He believed EHS was psychological, based on what he read from the

World Health Organization, as well as based on the electrosensitives' behavior. At one library meeting, Schou reacted frantically to the overhead lights, storming out and wandering around the parking lot, saying she was too disoriented to drive home. She requested that Johnson change all the lights to incandescent, which emit a fraction of the electromagnetic radiation produced by fluorescent lights, although the latter are more energy efficient.

"I said, 'You got the money to pay for that?'" Johnson recalled. "We were barely scraping by on a tight budget."

As a compromise, the Green Bank library—which was already the county's only library without WiFi—agreed to turn off its lights for the electrosensitives. Schou and other sensitives had high expectations for quiet. Locals had varying degrees of amenability. And Johnson was not alone in feeling Schou was trying to enforce "electromagnetic sharia law" on the county.

"You could write a murder mystery story about all this," he said. "Call it *Murder by WiFi.*"

IT'S HARD TO BE an outsider in Pocahontas. I heard myriad iterations of the phrase "If you haven't been here for six generations, you're still a newcomer." At first it sounded quaint, a note of pride in how families had settled the land. But it also hinted at how residents could be dismissive toward new people and new ideas, as if to say, *We don't care what you think because you're not from here.*

Diane Schou pushed the county's tolerance of outsiders to a new limit. I was told that at a square dance she flipped off the lights because they were bothering her, causing several elderly dancers to fall. (Schou told me she had no memory of this.) At the senior center, she unplugged the fish tank's air pump because it emitted electromagnetic radiation; the fish were later found dead. (Schou didn't deny this, and said other sensitives were also bothered by the air pump.) At a Christmas service, she reacted to the flash of a digital camera with violent sneezes and cries, writhing on the floor. (Schou confirmed this.)

For a time after she first arrived, Schou found acceptance at a Lutheran church, which gave up its wireless microphones, changed its heating system, banned cellphones, and began offering gluten-free communion wafers—all at her request. Then she started having trouble with the vehicles in the church parking lot, perhaps because people left cellphones in their cars. Newer vehicles also had gadgetry that affected her, Schou told me. The church said it couldn't dictate what vehicles people drove, so Schou switched to Wesley Chapel United Methodist Church, a mile from her home in Green Bank. It was a smaller church, which meant it had fewer cars, and even fewer after half of the congregation left in protest over Schou's request that the lights be kept off. Pastor David Fuller called for the church to be accepting of outsiders. He sympathized with Schou, his empathy informed in part by his father's death from cancer after working in a steel mill handling asbestos. Asbestos was largely unregulated in the United States before the 1970s. Maybe Schou was onto something now?

The community's frustration really erupted in 2010. Schou was a regular at the Green Bank senior center, which changed the lights in one area so she could feel more comfortable eating her daily subsidized lunch there. Then she requested that she be served lunch so she wouldn't have to walk beneath the fluorescent lighting in the self-serve area. During a meeting that she organized to share information on EHS, a senior citizen named Walter "Tony" Byrd stood and asked Schou why she demanded special treatment for a condition that was medically unproven. Byrd (a brother-in-law to Bob Sheets) himself was a diabetic with high cholesterol and high blood pressure, yet he didn't expect special foods or services. He learned that Schou had a history of filing legal complaints against organizations if they failed to accommodate electrosensitivity, and he would later show me a copy of the complaint that Schou filed against Northern Pocahontas County Health Clinic in Green Bank for "denying access" to an electrosensitive-safe bathroom. "This crap needs to stop," Byrd told the other seniors, to applause. The meeting grew heated, and the sheriff was called to defuse tensions.

At some point during all this, Schou found a woodchuck stuffed in her mailbox, shot to death. *Leave Green Bank,* was the message, *because you're not welcome here.* She reported the incident to the police but was met with a shrug. "I thought they could take the animal and analyze the bullet," Schou told me. "They didn't even come out." Another time, her car's tires were punctured in her own driveway. Other visitors to her home mysteriously had their tires slashed.

Schou stopped going to the senior center. When John Simmons took over the county's senior programming in 2010, he changed the building's lights back to the way they'd previously been. "Don't come back here in these hills and expect these old hillbilly people to change their ways," Simmons told me when I met him at his office in Marlinton. A wad of tobacco bulged under his lip. A longtime foreman at a leather tannery just north of Green Bank until it closed in 1994, Simmons was now responsible for organizing meals and services at senior centers throughout the aging county, as well as for delivering about one hundred meals to the elderly every day. He couldn't be bothered by a dubious illness unrecognized by the Americans with Disabilities Act or the American Medical Association.

"I don't know how many are in the county now," Simmons said of the sensitives. "They're like flies, they keep coming in."

To his point, it was impossible to get a solid count on the number of electrosensitives in Pocahontas. Some lived in the closet for fear of being ostracized.

"It's like modern leprosy," said one man who wished to remain anonymous. He had moved to Pocahontas from California with his wife and three kids in 2013, several years after he began experiencing brain fog, memory loss, and blurry vision around electronics. He tried everything to curb his symptoms. He cut all sugar from his diet. He exercised. He slept in special pajamas. He unplugged the electrical wiring in his car. He stopped drinking from aluminum cans because he believed they acted as antennas for radio waves. He considered becoming Amish. In 2013, he read about Green Bank and Diane Schou. He visited once and moved his whole family out. His son joked to schoolmates that his family was hiding in Pocahontas as part of a witness protection program; it was easier than explaining the truth.

"If it hadn't happened to me, I would for sure be super-skeptical of a lot of this stuff," the father told me. "I used to be one of those guys that got in line and waited for the next-generation of an iPhone."

But he wasn't interested in meeting Schou or being associated with the other sensitives. He just wanted to get better. He believed he could heal in the Quiet Zone and return to a seminormal life in the world of connectivity. The sensitives I spoke with all believed they recuperated in Green Bank, though they had varying degrees of tolerance for returning to the outside world. Schou, for her part, said the Quiet Zone made her feel strong enough to make periodic trips to visit family and friends, but she always had to return to Green Bank to recharge.

"I feel like they've given up on trying to be a part of the modern world," the man said of the sensitives who had settled permanently in the Quiet Zone, "and I don't want to."

"**HOW MANY SCIENTIFIC STUDIES** would you need to convince you that this is real?" asked Bert Schou, looking me in the eye solemnly. His wife, Diane, sat beside him at Station 2 restaurant in Durbin.

"A couple would be great," I said.

"This one will start you," Bert said, sliding a folder across the laminated tabletop with an air of gravity. He said it contained definitive scientific proof that electromagnetic hypersensitivity was real. In addition to the study in the folder, he claimed, some "twenty thousand" additional research papers showed concrete evidence of EHS.

We'd all just attended a Sunday morning service at Wesley Chapel United Methodist Church. Bert wore a collared shirt and slacks. Diane was in a flannel shirt and long skirt. I donned my sole dress shirt. By now I'd been visiting Green Bank for more than a year and had spoken with Diane many times, but this was my first conversation with her husband of forty years. Bert lived half the time in Iowa, where he ran an agricultural seed-testing company, but he intended to retire to Green Bank because he also suffered from "a little bit of sensitivity." Diane chimed in that Bert's hair thinned when he was working in Iowa, but it grew back "nice and thick" in Green Bank.

When I got back to my apartment at the observatory, I peered into the folder, skeptical of what I might find. I'd already received "evidence" from other sensitives. On one of my first visits, a woman named Jennifer Wood, whom I'd met at church with Schou, demanded that I sign a pledge to include a list of her preapproved "research" in whatever I wrote. When I declined, she accused me of being "pro-industry" and "paid off" by cellphone companies. "Who are you working for?!" she'd yelled, pointing a finger at me. After assuring her that I wasn't an industry spy, Wood handed me her "evidence" sheet. It said Russia had been aware of "radio wave sickness" for decades, but that fact was suppressed by the U.S. cell industry. Her sheet cited a 2016 study from the U.S. National Toxicology Program that showed male rats had a slightly higher rate of cancer when exposed to cellphones, though she left out that the male rats also lived longer and that female rats in the experiment showed no uptick in cancer—and none of the rats had cancer rates statistically higher than the average rat population. Wood's sheet also highlighted how the World Health Organization in 2011 classified electromagnetic fields as "possibly carcinogenic to humans (Group 2B), based on an increased risk for glioma, a malignant type of brain cancer, associated with wireless phone use." Thing is, drinking very hot beverages like coffee is considered even more dangerous and "probably carcinogenic to humans" (Group 2A), but I've never seen a cancer warning on a Starbucks cup. Moreover, the World Health Organization has since 2005 declared that "there is no scientific basis to link EHS symptoms to EMF exposure."

In the Schous' folder was a 1991 study called "Electromagnetic Field Sensitivity." The lead author was Dr. William Rea, founder of the Environmental Health Center in Dallas, where Wood had spent time. In the study, Rea exposed one hundred self-diagnosed electrosensitives to frequencies at random intervals to test their ability to sense radio waves. One-quarter of subjects reacted consistently to stimulation; sixteen showed a hyperawareness of when they were exposed to a certain frequency. "We concluded that this study gives strong evidence that electromagnetic field sensitivity exists, and can be elicited under environmentally controlled conditions," Rea and his coauthors wrote.

I looked up Rea's name. He'd been charged by the Texas Medical Board in 2007 with using pseudoscientific testing methods, failing to make accurate diagnoses, and providing "nonsensical" treatments. There was also a history of malpractice complaints against him. So much for a reliable source.

The electrosensitives recommended that I read the book *Overpowered,* by Martin Blank, a Columbia University Ph.D. I did, and I was astounded by its dubious claims, including that metal eyeglass frames focus cellphone waves "directly into your brain" and that living within close range of a cell tower increases one's risk of suicide. I was also directed toward an activist in New Mexico named Arthur Firstenberg who once sued his neighbor for $1.43 million for having WiFi that allegedly damaged his health. Another "expert" the sensitives mentioned was David Carpenter of the State University of New York in Albany. Carpenter had earned his medical degree from Harvard, which lent a degree of authority to his assertions that EHS is real and that 5G cell service is harmful—a debunked conspiracy. In the journal *Child Development,* professors from Oxford and Queen's University Belfast found a number of Carpenter's claims to be "scientifically discredited" and "widely dismissed by scientific bodies the world over."

In a major paper that poured cold water on the very concept of EHS, James Rubin of King's College London reviewed forty-six blind or double-blind provocation studies involving more than one thousand test subject volunteers. He found no "robust evidence" supporting the idea of EHS, according to a study published in 2010 in the peer-reviewed journal *Bioelectromagnetics.* According to Rubin, in single-blind tests when the researcher knew that the subject was being exposed to an electromagnetic field, people *did* consistently experience EHS symptoms. But when neither the researcher nor the subject knew if the electrical field was on or off—that is, when the experiment was performed double-blind—symptoms disappeared.

Rubin attributed electromagnetic sensitivity to the nocebo effect, which is when you feel physically unwell after you think you've been exposed to something hazardous. He ran an experiment where people watched a video about the harms of WiFi and were then exposed to a fake WiFi signal. Those who watched the video were more likely to report feeling pain from WiFi (even though there was no actual WiFi).

I witnessed the nocebo effect with many of the electrosensitives whom I met, and most obviously with Schou. She complained about a neighbor's WiFi and claimed to be able to attend Wesley Chapel only when the lights and furnace were off, but she had no problem eating at Station 2, which had bright incandescent lights and strong enough WiFi that the sheriff's deputies sometimes loitered outside to access the free hotspot. When Schou and I attended a meeting of the county's amateur radio club, she insisted on sitting in a dark corner lest the overhead lights bother her, even though the entire building had WiFi. She said she could eat eggs and poultry in Green Bank, but if she consumed the foods outside of the Quiet Zone they gave her "explosive diarrhea." She said she was sensitive to electric coffeemakers, yet I saw a Keurig in her house. She said she had trouble driving beneath power lines, yet she regularly drove to Baltimore to visit her son and his family. She was allergic to microwaves made by Amana, but not microwaves made by Sharp or General Electric. She said she had type 2 diabetes and her blood sugar spiked when exposed to electromagnetic radiation, yet she and Bert had an iPad.

Perhaps most peculiar, Schou claimed to have an intolerance for "rippling brooks," perhaps because she once stayed in a cabin in Sweden by a waterfall that supposedly faced in the direction of Ukraine and had somehow "absorbed" radiation from the Chernobyl nuclear accident. Ever since then, "If I am near a rippling brook, some people who are electrosensitive cannot be around me then because they can detect something from me," Schou said. "If I go to visit them from a different direction, not parallel to a rippling brook, they don't have a problem. But if I go to see them via the way of a rippling brook, they cannot be around me. It is bizarre."

SCHOU WAS A MEMBER of the Deer Creek Valley book club, which I joined whenever possible. The club was led by a woman named Carla Beaudet, who was an engineer at the observatory. She had a degree in electrical and computer engineering from Johns Hopkins University and had previously worked as a radio frequency compliance tester for commercial products. Because of her knowledge of safety protocols, and because of her interactions with Schou, she seemed well positioned to offer an objective view on the potential dangers of electromagnetic radiation.

Beaudet could often be found in a rear annex of the observatory's science building, conducting tests inside something called the anechoic chamber, a rectangular room wrapped in a double layer of galvanized steel that insulated it against electromagnetic fields and made it a place of radio silence. Inside, the floor, walls, and ceiling were covered in blue foam cones made from conductive carbon. At one end of the thirty-seven-foot-long chamber, an antenna could capture electromagnetic radiation emitted from whatever gadget Beaudet placed at the other end, allowing her to test its noise level. Usually she was reviewing equipment to be installed on the telescopes, but she also helped keep watch for intrusive gadgetry. One of her experiments led to Fitbits being banned at the observatory.

After showing me inside the chamber, Beaudet invited me back to her office to talk more. As we walked through a hallway, she pointed to a ceiling-mounted security camera. It had a mesh screen on the inside that blocked the electromagnetic radiation from leaking out, something she'd designed. She'd built similar Faraday cages for everything from the motors on the Green Bank Telescope to the LED lights overhead. Another project was outfitting the neighboring public school's outdoor digital sign so it wouldn't emit electromagnetic radiation, a project that required no fewer than nineteen tiny Faraday enclosures around the internal electronics.

In Beaudet's office, she took a moment to feed her pet African clawed frogs, sprinkling some chicken liver into their glass tank. On the wall, a climbing harness hung ready for when she needed to climb the telescopes to work on equipment. She boiled water for tea. Since moving to Green Bank from Baltimore a decade earlier, she had come to appreciate the area and its culture. She gardened, foraged, and bowhunted, with a goal of one day shooting a deer from her porch. She had more friends in this sparsely populated area than she'd had in the city, in part because Green Bankers couldn't afford to not be friends. "You can't get away with being an asshole for very long, because then you're really isolated," she said. "The interactions here are more valued." In a way, Beaudet had also come to believe in the healing powers of the Quiet Zone, with a perspective that shed light on why Schou and other electrosensitives might feel better in Green Bank.

"When I drive here from D.C., as the traffic goes away and the landscape becomes greener and quieter and nicer, I relax," Beaudet said. "I start feeling better. I ascribe this to visual, auditory, olfactory—all the senses that I'm aware of. It's just nicer out here. It's gentler . . . Being out of an overstimulating environment helps us relax, keeps us healthier, all of that."

But that didn't mean she believed WiFi or cell service was harmful to humans, much less that a person could physically sense those radio waves.

"I honestly don't believe in [electromagnetic sensitivity], and that's because the people who have it have such a vast array of symptoms," Beaudet said. "We're talking about anything from headaches to cancer. So many different symptoms are blamed on this phenomenon. You can always find a study here or there in support of whatever you want, but more studies show no compelling connection."

EVEN DIANE SCHOU was skeptical of some electrosensitives. One long-term visitor to her home refused to help clean because he worried he'd lose his disability paycheck if someone saw him doing manual labor. Another requested a specific wine paired with every meal. "Some people want to come and I get the feeling that they're not really harmed strongly by electromagnetic radiation," Schou said. "They're looking for somebody to take care of them."

A number of electrosensitives exhibited signs of paranoia, depression, hypochondria, and outright kookiness, saying they were being followed by drones or monitored by the CIA. One told me he could electrify lightbulbs with his bare hands. Another claimed he'd visited Roswell, though I'm unsure what UFOs had to do with EHS. A woman believed she could hear music from rainbows. I met a Green Bank farmer who had become convinced that EHS was real, or at least as real as chemtrails—mind-control particles in the form of jet engine vapor that the government supposedly sprays over America to keep citizens docile. The farmer also believed Barack Obama was born outside the United States, that Hillary Clinton murdered a staffer, that John F. Kennedy was assassinated by his government, that "for years" there had been a cure for cancer, and that vaccines cause autism—all of it apparently masterminded by an evil cabal in Washington, D.C.

I had thought the Quiet Zone might be spared from such conspiracies because it was mostly offline and distanced from the inanities flying around the web. Instead, I was finding the same delusions as elsewhere. Isolated in Green Bank, people could retreat into their own minds. They heard whatever voices they wanted to hear. Wild ideas festered.

"There's people that come here with mental illnesses who are looking for answers," said Green Bank resident Sue Howard. "It's a magnet for weirdos."

Howard had a unique perspective on the matter, as she herself was an electrosensitive. In her mid-fifties, lanky with blond hair, she had moved to Green Bank in mid-2016 from Westchester, New York. She lived in a mobile home that Schou rented out for $400 a month. Her porch had a view toward the Green Bank Telescope popping above the trees a few miles away.

Of all the sensitives I was meeting, Howard seemed the most relatable, perhaps because we both came from the New York area. She had felt sensitive to electromagnetic radiation since around 2009, when touching a computer mouse caused a tingling in her hand. Texting on a cellphone soon caused pain and numbness up her arm. Being near high-power cell antennas gave her "a sharp stabbing pain" in her head where she used to wear a metal hair clip. In 2011, she was diagnosed by an infectious disease specialist with "severe progressive sensitivity to electrical fields." Like other electrosensitives I met, she suffered from multiple chronic illnesses, including chemical sensitivity. When she read about Green Bank in a 2015 article that mentioned Schou, Howard immediately began plotting a way to get there.

She first visited Green Bank the fall of 2015 with her husband, camping at the fourteen-acre property owned by WAVR, Schou's nonprofit. Driving down a dirt road to the campsite, the Howards' headlights had shone upon two rotting cow skulls hanging from spikes on a neighbor's tree. It was like a demonic warning against venturing deeper into the forest. But there was no turning back. They had journeyed five hundred miles to an area where they knew nobody except Schou, who was already proving eccentric. She'd asked the Howards to bring Pillsbury gluten-free pie dough—a rare commodity in Pocahontas—but they had found only gluten-free pizza dough. "It's not pie dough," Schou had said, oozing disappointment.

The Howards pitched a tent in the dark. It poured overnight, turning the ground into a muddy soup. Still, Sue awoke to an incredible feeling. For the first time in years, she felt free from pain. On an impulse, the Howards purchased a thirty-eight-acre parcel of undeveloped land where they planned to eventually settle. For the time being, they couldn't stay. Sue's husband, John, was tied to a job in New York, where they owned a home with his aunt.

Back in Westchester, Howard felt increasingly sick, and at her wit's end. She'd already hired a "building biologist" who had inspected their home and recommended they change their electrical wiring and paint their house with something called YSHIELD that claimed to block electric fields. Howard had also relegated the family computer to the basement, banished her son's Xbox, and removed the smart meter and WiFi from their home. They'd changed their lightbulbs to incandescent, but Howard preferred the lights simply off. Around 2015, she'd plastered a one-hundred-square-foot room of her home with special silver wallpaper to block out electromagnetic radiation. Her family called it "the silver room." The Quiet Zone sounded like a community-size "silver room."

By May of 2016, Howard decided she could no longer tolerate New York. She moved alone to Appalachia, leaving her husband behind in Westchester. (Their children, both in their twenties, had already moved out.) Living in Green Bank, her vision improved, her tinnitus went away, and a heart arrhythmia disappeared, though she said it returned when she visited nearby cities. She no longer had to wear an EMF-shielding scarf or a face mask for chemical sensitivity, but she still took other precautions. She operated her home's washer, dryer, and refrigerator only when she went out, and she often kept the electrical breakers off altogether.

Her husband tried to visit every couple weeks, bringing organic groceries and VHS movies because she couldn't tolerate the radiation from DVD players. Her son and daughter also visited, though they were less enthralled by the quiet. One weekend, her son drove twice to Snowshoe Mountain Resort, an hour away round trip, for a fix of cell service and WiFi. Howard, for her part, checked her email regularly at the Green Bank library. "The librarian is wonderful and she's willing to turn off the lights for me," she said.

Howard picked blueberries, jumped into a swimming hole, tasted bear meat, and gained a new sense of self-confidence from all the media reports about the growing electrosensitive community around Green Bank. She was interviewed by Danish filmmakers, by a Russian reporter from RT Documentary (part of the Russian TV network), and by a Netflix crew for a documentary series called *Afflicted* about mysterious illnesses. She had felt invisible in New York, only to find stardom in Appalachia.

"I walked into Hollywood here," Howard said. "Who knew?"

But the Quiet Zone wasn't perfect. The rural electric lines were staticky. Neighbors had WiFi. People carried smartphones. And she'd walked into a thicket of tensions caused by the behavior of some electrosensitives, which forced her to tread lightly.

Then there was me. When I turned on my iPod to record our conversation, Howard eyed the device warily and asked that I keep it at a distance. Minutes later, she jumped.

"Look!" she said. "I just got a pop from that. See what happened to my vein right now? It just went *pow*."

My iPod, she said, had caused her vein to pulsate and "pop." I said I couldn't see any change. Her veins looked just like mine.

"It wasn't like that a minute ago," she insisted.

Howard took out a radio frequency power meter and looked with concern at the voltage reading. "This is too high for here," she said. "I knew something was off."

ACCORDING TO a "cultural resources evaluation" prepared in May 2016 as part of the National Science Foundation's review of whether to continue funding the Green Bank Observatory, a vulnerable group that would be affected by the facility's potential closure was "individuals seeking to avoid health effects that they perceive from electromagnetic radiation who have chosen to live in the NRQZ as a 'safe haven' from that radiation." Sue Howard, among others, told me she was "working closely" with the observatory to help protect the Quiet Zone. Sensitives regularly contacted business manager Michael Holstine to report when they felt pain, asking him to look into it for their health and for the good of radio astronomy.

One time, Howard reported to Holstine that she "sensed" something awry in Green Bank. According to Howard, Holstine confirmed to her that a military communications satellite system known as MUOS had been orbiting overhead when her sensitivity was triggered. It sounded like definitive proof that she could sense electromagnetic radiation like a kind of superpower.

I later checked this story with Holstine. He said Howard had indeed called him to say she felt something, and he had mentioned that communications satellites orbited over Green Bank. But he denied that he'd established any correlation between her sensitivity and a particular satellite overhead. There was simply no way for him to know at what precise moment a satellite might be transmitting toward Green Bank. Nor was there anything he could do about it, as satellites were outside the jurisdiction of the National Radio Quiet Zone.

"Did you detect *anything* that she reported?" I asked.

"No," Holstine said flatly. But did it matter in the end? "All I know is they feel better when they're here. They're the people who are *not* going to use electronic devices. So, for the observatory's sake, they're great neighbors."

CHAPTER THIRTEEN

"Papers and Pencils"

BANG!

"I don't know where that went," said seven-year-old Coleton Birely, lowering his .22-gauge shotgun. He'd recently gotten his first firearm, a bit late by local standards.

Bang!

"I swear that was right on target!" Coleton said, puzzled as to why both his shots completely missed the target sheet in his yard.

I'd first met Coleton at Trent's General Store, finding him seated on a checkout counter doing homework. His mom, Debbie had been working the register. They'd moved to the area in 2012 from Spokane, Washington, in part because Debbie had relatives in Durbin. She'd also been looking for a new start. Pocahontas was welcoming, with its slower pace and easygoing social interactions. There was also a sense of fate in moving to Appalachia. Debbie couldn't recall how she'd first heard the name Coleton, as it was uncommon in Washington, but in West Virginia she'd call for her son and a handful of Coletons would look up. There was even a town called Coalton. It felt like a sign they were meant to be there.

To Coleton's house, I'd come equipped with ammunition and safety goggles purchased from Trent's. Inside on the refrigerator,

I'd spotted a coloring activity of a cartoon figure with the caption: "My dad's pockets are full of _____." Coleton had inserted "phone." (Coleton's "dad" was Debbie's boyfriend.)

Coleton again took aim at the target sheet nailed to a tree about twenty feet away. He'd been taking shooting lessons at the Green Bank Observatory's recreation fields along with swim lessons in its underground pool—both ways that the facility opened itself up to the community.

"Remember your breathing," Debbie coached, standing by in gray sweatpants and a black hoodie that framed her bleach-blond hair. "Inhale and shoot on the exhale."

"What kind of animals could you shoot with that gun?" I asked.

"Squirrels, deer," said Debbie. "If someone were to accidentally get shot with that, it would put you in the hospital and possibly kill you."

I stepped back. Coleton was living a full life in many ways. But was it worthwhile for no cell service and halting internet? Sure, he'd learn how to shoot a gun, dress a deer, drive a stick, use a compass. But would he be able to function outside the Quiet Zone? Debbie had concerns, especially around education. Coleton was attending Green Bank Elementary-Middle School, which fell in the shadow of the radio telescopes, meaning it had the only classrooms in the entire country where WiFi and iPads were essentially outlawed.

Coleton passed me another gun, this one a heavy .38 double action revolver.

"It's got a bit of a kick, so watch out for it," Debbie warned, as I squared up to the target and squinted into the aiming sight. Last time I'd held a gun was in Cambodia, nearly a decade earlier, at a shooting range that gave tourists the chance to fire bazookas and hurl grenades.

Bang!

"By Virginia, he hit the target!" Debbie exclaimed. "Did you see the flash?"

"I think my eyes were closed," I said.

THE FIRST TIME I walked into Coleton's school, a student was in trouble for creating an illegal hotspot using a teacher's computer, arguably breaking state law in broadcasting an unapproved radio frequency at the center of the Quiet Zone. I got the impression it wasn't the student's first offense.

I found the culprit, an eighth grader named David Bond, in a cinder block–walled room strewn with musical instruments. He readily confessed to knowing how to create a hotspot using school computers as well as how to sneak around the administration's web filters to access social media. Wearing Beats by Dre headphones and a "Macho Man" Randy Savage tank top, he said he owned both an iPod Touch and an iPhone 6S. He added that most of his classmates also had smartphones. His music teacher, Greg Morgan, agreed—and showed me his own smartphone.

"Some of these kids are very poor, but they still have an iPhone 7 or whatever," Morgan said. Between 30 and 40 percent of children in the county lived in poverty. Every student received free breakfast and lunch through the Healthy, Hunger-Free Kids Act. But that didn't limit the pervasiveness of smartphones. When I later asked a class of third graders if anyone owned a smartphone, every hand shot up. Even in that room of eight- and nine-year-olds, I appeared to be the only person without a cellphone.

Aside from creating a hotspot, Bond was also in trouble that day for passing a note to his girlfriend. "It was just something sweet," he said of it. (And I'd thought handwritten love letters were obsolete.) The young couple also communicated using their iPhones' AirDrop function and on Snapchat and Facebook, which Bond checked once a day on his home internet, though he didn't really consider it internet because the speed was so slow—0.28 megabits per second upload and 0.09 megabits download, compared with the national average of 64 megabits upload and 23 megabits download. Video streaming was impossible. The speed had nothing to do with the observatory and was hardly unique to the Quiet Zone, though it was symptomatic of living in a remote, sparsely populated, quiet place. An estimated one-third of all people in rural America have little or no access to the internet. West Virginia ranks close to last in the nation for broadband penetration.

Internet was much faster at school thanks to a dedicated broadband connection, though everything had to be wired by order of the observatory, which presented its own difficulties. Morgan couldn't use the WiFi features of his SMART Board, which would have allowed him to control it remotely with an iPad and more easily roam the classroom to interact with students. At school events, Morgan had to find workarounds when guests showed up with wireless mic systems. If the speaker wanted to walk into the assembly to interact with students, Morgan followed behind with the microphone cord as it snaked through the aisles, tangled on chair legs, and draped across students' laps.

And without the convenience of turning on a WiFi router, it was more logistically complicated and expensive to expand internet to every classroom and provide web access to every student. For online testing, Green Bank students had to rotate through the school's two wired computer labs, which could take weeks. (Other schools in Pocahontas County were far enough from the telescopes to have WiFi and utilize mobile computer labs.) It was impossible for administrators in Green Bank to do mobile observations over an iPad or iPhone camera, or to enter teacher and student evaluations directly into a WiFi-connected laptop.

The educational hurdles extended outside the school. An estimated half of all students and teachers lacked fast enough home internet to do online learning or take advantage of streaming programs, according to Ruth Bland, director of technology for the county's schools. Whereas people in other areas of the country with slow internet could fall back on cell service for online connections, that wasn't an option in most of Pocahontas, making it harder for students to remotely access libraries and educational websites such as Desmos and Khan Academy.

"Folks living on the edge get their internet through cellphones, so if we can't have cellphones, then we are impacting people's connectivity," said Joanna Burt-Kinderman, the county math coach (and sister of Sarah Riley, director of High Rocks). "If you don't have internet speeds that allow students to use free, open-source, online tools, that's a big limitation in an age when education funding is getting cut."

Bond was less concerned about cell service and internet for education than he was about downloading software updates for his Xbox. To do so, he had to drive an hour north to the city of Elkins, where his best chance for fast internet was tethering to the WiFi at Sheetz, a souped-up gas station and convenience store chain. The county's slow internet was a reason Bond wanted to move away as soon as he could, though he said he'd miss hunting in the area. He started naming all his guns: "I've got a .45 ACP, a 7mm-08, .243, .17 HMR, a 12-gauge, 20-gauge, .30-30, .22, a .410, another .243, muzzleloader, and a .380 pistol."

"How many guns is that?" I asked.

"Thirteen, I think. Hold on." He recounted. "Twelve."

Bond shot his first deer at age eleven, the age most kids in America got a smartphone. The *Pocahontas Times* regularly published photos of children beside their first kill.

"I've got videos of me shooting a gun," he said. "You want to see one?"

"Sure," I said.

He whipped out his iPod Touch.

FORTUNATELY FOR BOND, just as he entered Pocahontas County High School in the fall of 2017, the administration lifted its ban on mobile devices. Previously, if a student took a cellphone out of their locker, the device was confiscated and the student's parents were contacted; on second offense, the student received in-school suspension. Now, teachers had discretion over smartphone use in their classrooms.

The policy change was primarily because students had become inseparable from their devices, even if there was no cell signal or public WiFi at the high school. Within a year of the policy change, Principal Joseph Riley estimated that three in four students were carrying smartphones. Math teacher Laurel Dilley put the figure higher. "All of them," she said without hesitation, "unless they're really unfortunate."

One morning, I swung by Dilley's classroom to gauge the situation for myself. The white walls were decorated with math and computer science posters. One said "#codeislife." Another read "No Electronic Devices Allowed," a holdover from the smartphone ban. Dilley had previously taught in Morgantown, the state's third-largest city, where she'd witnessed students texting each other test answers. That kind of cheating was less possible in the Quiet Zone, though it was getting easier with the influx of smartphones that could communicate via Bluetooth. Glancing around Dilley's classroom, I saw that most of the eighteen students had smartphones on their desks. I asked them what was the point of carrying cellphones in a Quiet Zone.

"I can use it in Lewisburg," one explained. "I go there once a week."

"Is it worth paying for a smartphone data plan to use it once a week?" I asked.

"Yes!" a bunch of students shouted.

While the school's WiFi was only available to administrators, students regularly hacked in and traded the password like a commodity. A teacher said she'd seen students pay up to twenty dollars to get the code, which provided them with a temporary internet fix until administrators reset the password.

"Does anyone here live in Green Bank?" I asked the class.

Three hands went up.

"Do you have smartphones and WiFi at your homes?"

All said yes. They either didn't realize or didn't care that they were potentially admitting to breaking state law.

"Does anyone know why cell service is limited in Pocahontas?"

"Because of the big TV dish."

"If you had to choose between cell service and the observatory, what would it be?" I asked.

"Cell service!" a student yelled. "Send that sucker overseas!"

Staff had also embraced smartphones. A physics teacher had students use their smartphones' flashlight function to observe how light reacted to polarization filters. The forensics class used the smartphone as an audio recorder. Math teachers appreciated the smartphone's capability as a graphing calculator, as many school-provided devices were broken. The assistant principal and athletic director, Kristy Tritapoe, told me she was constantly on her iPhone between WiFi at school and WiFi at home. "When I wake up at three o'clock in the morning, I check my phone," she said. "I'm checking my email, checking Facebook—I'm just checking it." So much for there being one place in America where you left your work at work.

I met one student who seemed to share my sentiment toward smartphones: Mathias Solliday, the young man whom I'd run into at Trent's (and whose Green Bank home not so secretly had WiFi). He didn't own a cellphone and didn't see the need for one. He said he was happy with his iPod.

A typical high schooler in many ways, Mathias ran on the track team, was an officer in Future Farmers of America, and competed with the elite forestry club, which has won eight national championships and twenty-six state championships since 1990. The club had free rein over eighteen acres of wooded school property that abutted the 11,684-acre Seneca State Forest, where Solliday was fine-tuning his skills in agronomy and timber management. Forestry competitions entailed identifying tree species and insects as well as cruising the woods using an old-fashioned compass.

For one forestry project, Solliday and his classmates teamed up with the observatory to analyze the long-term health of the tall pines planted around the telescopes to act as a natural barrier against radio noise. The students developed a twenty-five-page management plan that recommended thinning old growth and planting new seedlings. They cut, milled, and kilned several hundred trees and sold the wood to a local lumber company, using the funds to purchase a new van for traveling to competitions. It was real-world forestry that would have been inaccessible to most schools nationwide.

Not that Solliday was only proficient in tromping through the woods. He was also in the school's STEM club. In early 2017, he and several other members won a statewide competition for a smartphone app proposal to use weather data to warn residents about flash flooding. (The previous summer, twenty-three people had died in floods across West Virginia when ten inches of rain fell over twelve hours.) Their app proposal won first out of 1,800 entries in the 2017 Verizon Innovative Learning app challenge. Solliday personally received a Verizon tablet, and the STEM club got $5,000 in prize money that it used to purchase a 3-D printer.

"We live in this area with no cellphone service, horrible internet, and we were number one in the state," Solliday said. "I don't feel like we're at a disadvantage at all."

Solliday's team never actually developed the app. Word of mouth and landline calls would remain the primary ways of warning about flash floods.

"If somewhere is flooding, we're going to get called," he said. "Everybody is connected in Green Bank, everybody knows where everybody lives." His entire house rang to life when anyone called; it had five wired telephones. "Say a cow is loose, we'll go to that person's house and tell them the cow is out and help them put the cow back into the field."

In the end, Solliday's stance against smartphones was short-lived. By his senior year, he was also carrying around an iPhone.

WHILE SMARTPHONES WERE infiltrating Pocahontas, many schools outside the Quiet Zone were instituting phone bans to cut down on online distractions—in essence, trying to make schools *more* quiet amid studies showing negative side effects from smartphones. The devices cause a "brain drain," diminishing "learning, logical reasoning, abstract thought, problem solving, and creativity," according to a 2017 study in the *Journal of the Association for Consumer Research.*

Public school test scores fell in Baltimore after the city adopted a one-laptop-per-child policy. Banning phones in school has been shown to lead to higher student test scores. A survey of ninety-one schools in England from 2001 to 2013 found that classrooms that banned cellphones on average saw a nearly 6.5 percentage point boost in test scores. Low-achieving students' scores rose 14 percentage points. No single study is definitive, but there have been enough to cause concern.

Amid research showing ill effects from too much screen time, Taiwan in 2015 outlawed tablets and other electronic gadgets for all children under the age of two, with a $1,500 fine for rule-breaking parents. All Taiwanese under the age of eighteen were ordered not to use digital media for "a period of time that is not reasonable." In 2015, France banned WiFi in day care centers and established regulations on WiFi and cell towers throughout the country, including mandating that all WiFi hotspots be clearly labeled. France later extended its ban on smartphones through ninth grade. In the United States, the American Academy of Pediatrics in 2016 recommended that children under eighteen months avoid all screens other than video chatting and that children aged eighteen to twenty-four months watch only "high-quality programming" with their parents. The World Health Organization in 2019 issued guidelines of no screen time for children under two and no more than one hour of screen time per day for children ages three to four. The same year, the Canadian province of Ontario banned cellphones in classrooms.

The alarm bells are even ringing in Silicon Valley, of all places. "I am convinced the devil lives in our phones and is wreaking havoc on our children," Athena Chavarria, a former executive assistant at Facebook, told the *New York Times* in 2018. She did not allow her kids to have cellphones until high school. "Facebook is a fundamentally addictive product that is designed to capture as much of your attention as possible without any regard for the consequences," Sandy Parakilas, the company's former platform operations manager, told *New York* magazine in 2018. "Tech addiction has a negative impact on your health, and on your children's health." The same year, Chamath Palihapitiya, Facebook's former vice president of user growth, was quoted in the *New Yorker* saying Facebook was "destroying how society works—no civil discourse, no coöperation, misinformation, mistruth." Palihapitiya said his children were "not allowed to use this shit." The Twitter engineer who invented "pull-to-refresh" later repented for the feature's addictive nature. The Facebook engineer who created the "like" button had a parental control set up on his own phone to stop him from downloading apps. Tristan Harris, former design ethicist for Google, left the online search giant in 2015 to focus on what he called the Time Well Spent movement, which led to his founding the Center for Humane Technology to combat the "digital attention crisis." For Pocahontas, the lack of cell service and restrictions on WiFi were arguably blessings in disguise, allowing for distraction-free spaces that could aid concentration.

Some parents actually transferred their children *to* Green Bank Elementary-Middle School because of its proximity to the observatory. Gayle Boyette switched her son from Marlinton to Green Bank because it had a greater diversity of students and educators, owing to the influence of the observatory's staff. He joined the school's robotics club, which was coached by the observatory's director, an astrophysicist, and her husband, a software engineer. In 2018, the robotics club placed first in the West Virginia First Lego League tournament, earning a world championship berth. Another software engineer from the observatory helped teach the high school's college-credit coding class and coach its robotics club, which in 2020 would place second in a statewide competition.

Despite making Green Bank seem behind in terms of wireless

technology, the observatory brought cutting-edge science into Appalachia and made it accessible to youths. Every year, an estimated five thousand students from around the region toured the telescopes, many for multiday immersion seminars. Boy Scouts came to earn merit badges in radio astronomy. The observatory hosted the high school prom and science fair, gave an award every year to a graduating senior, and sent its staff to assist in the schools whenever possible. When Green Bank Elementary-Middle School got wired internet, the observatory hooked it up. When the high school needed a new sound system, an observatory engineer helped install it.

Sam Felton, who was a Methodist minister along with being the mayor of Marlinton, praised the observatory's educational influence. He recalled an eye-opening field trip to see the telescopes when he was in high school. "Talk about broadening your horizons, that's interesting when you realize there's something going on in the heavens that we are connected to and have a desire to know more about," he told me. "We need to look up from time to time. King David said, 'When I look upon sun, moon, and stars, what is man that thou art mindful of him?'" Felton didn't have a problem with the astronomers dating the universe back billions of years, even if it contradicted a creationist view of the universe forming in six days about ten thousand years ago (a view held by two in five people in America).

I found Felton's perspective refreshing. The observatory had clearly changed local attitudes toward science. It even helped turn local kids into astronomers. Hundreds of high schoolers had interned at the observatory over the decades. A scientist named Ron Maddalena told me he'd overseen the internships of more than forty students, some of whom grew up in log cabins without electricity or running water—or, in one young woman's case, on a haunted mountain at the southern end of Pocahontas County.

HANNA SMITH HAD been interested in astronomy ever since her father gave her a *National Geographic* poster of the solar system. Her mind was further set on space when a middle school teacher showed her the documentary series *Cosmos,* hosted by Carl Sagan. She developed an intense crush on Sagan and built a shrine to the famous cosmologist in her bedroom, taping newspaper clippings about him all over one wall, with his books lined up beneath. She didn't learn until later that Sagan had visited her county in 1961 to discuss the potential for alien communications.

Hanna's home didn't have a computer, much less internet or cell service. Her family lived in a rickety cabin atop Droop Mountain Battlefield State Park, where her father, Mike Smith, was the park superintendent for three decades. Their house was originally built as a toolshed in the 1930s by the Civilian Conservation Corps. They were the only people living in the nearly three-hundred-acre park, aside from the ghosts.

About four hundred soldiers died on Droop in the Civil War, and their spirits were said to haunt the surrounding woods. When Hanna's uncle visited, he awoke in the middle of the night to the sound of horses jumping over a fence and coming up to the window. He refused to sleep there again. A past superintendent's son once heard what sounded like a horse clattering down the road, but he could see no horse; he then felt a horse neighing into his face. Years later as an adult, he told Hanna's dad that he still had nightmares about that horse. According to the county's history book, two girls once found two battle rifles in the park and carried them home, only to be pelted with rocks flying from all directions until they returned the guns to Droop. So pervasive was the belief that Droop was haunted, I was told that locals opposed a cell antenna in the park because it might disturb the dead.

Along with ghosts, Hanna was surrounded by some of the darkest skies on the East Coast. In 1996, when sixteen years old, she and her dad camped outside so they could spot the comet Hyakutake overhead. The Green Bank Observatory was also monitoring Hyakutake that night, detecting ammonia and water coming off the comet's tail. "The comet was real dim, but Mars that night was really bright," her dad recalled. "She decided then and there that she was going to be the first woman on Mars."

In high school, Hanna interned at the observatory. One of her jobs was to transfer floppy disk files onto compact discs—a tedious task, but still a unique opportunity for a high schooler to be peripherally involved with cutting-edge astronomy. At the end of Hanna's internship, Ron Maddalena wrote her a recommendation for Smith College in Massachusetts, one of the top liberal arts schools in the country. Tuition was more than what Hanna's father earned in a year, but she qualified for financial aid. She would finish in the top 1 percent of her class, summa cum laude.

Through college, Hanna dated her high school sweetheart, Nathaniel "Dane" Sizemore, who was majoring in computer science at Westminster College in Pennsylvania. They'd first met in sixth grade, when Dane bested Hanna in the county's annual math competition. (The *Pocahontas Times* photographed Hanna standing beside Dane with his first place trophy, looking like a stand-in for the character Will Byers in *Stranger Things*.) In high school, they'd found they both loved *The Hitchhiker's Guide to the Galaxy* and were interested in astronomy. Dane's father, Wesley Sizemore, was the observatory's Quiet Zone cop, and Dane had occasionally joined in the hunt for radio frequency interference. Dane recalled a particularly troublesome electric pole by Green Bank Elementary-Middle School that had to be pounded every few years with a sledgehammer to rattle the connections and stop it from arcing. He and Hanna attended the junior and senior proms together and got in trouble for having their hands around each other's waists, deemed too much physical contact.

They graduated from college in 2001, got married atop Droop Mountain, and moved to Colorado so Hanna could start a Ph.D. program in astrophysical and planetary sciences at the University of Colorado, Boulder, with a focus on Mars. The first six months in Boulder, they slept on an apartment floor in sleeping bags that were a wedding gift from Hanna's dad. At the end of the doctoral program, Hanna participated in NASA's Phoenix mission to Mars, in which a robotic spacecraft landed for the first time in the red planet's polar region to look for ice. Her research helped determine where the spacecraft landed, and the findings led her to author a number of papers analyzing the Martian landscape for ice and permafrost.

In 2009, they moved to California so Hanna could pursue a postdoctorate at NASA's Ames Research Center in Mountain View. Dane worked in IT at Google's campus. It suddenly felt possible for two people who had grown up in the disconnected Appalachian Mountains to make it in Silicon Valley.

The same year, they had twin boys. Soon they were paying about $30,000 a year for childcare, plus another $36,000 for a two-bedroom apartment. It felt unsustainable. In 2010, they returned to Pocahontas to be closer to family. Both wondered if their careers had been derailed.

Dane took a job with the state's office of technology and later joined the Green Bank Observatory as a software engineer. Hanna became an adjunct scientist at the observatory, which provided her with an office and fast internet that allowed her to do contract work for NASA and the Planetary Science Institute. In 2018, she made national news for helping discover "cryovolcanoes" on the dwarf planet Ceres. In support of NASA's Mars Exploration Program, she was also part of a multiyear effort to map subsurface ice on Mars, identify potential water sources, search for signs of life, and establish landing sites for a future human mission. If Hanna couldn't be the first woman on Mars, she would at least help pave the way for others.

Once young math rivals, the Sizemores now found themselves returning to the county's annual math field day as parents, encouraging students to integrate math into their careers. Dane would say, "Don't let anybody tell you that you can't do anything you want. Anybody from any little town or farm in Pocahontas County can go off and be a doctor or lawyer or businessman." Or a Mars scientist. Hanna would show a slideshow of mathematicians, scientists, and engineers from West Virginia, telling the students, "Look at these luminaries who are from here!"

THE SIZEMORES SENT their twin boys to Green Bank Elementary-Middle School, which fell into an ambiguous area with the Quiet Zone regulations. The administration wanted to be a good neighbor to the observatory and not emit radio interference toward the telescopes looming beyond the playground. But WiFi was already so pervasive in town that it arguably didn't matter if the school installed it. Along with an educational argument for WiFi, sometimes the technology was medically imperative. For a time, the observatory permitted the school to broadcast a weak WiFi signal for a diabetic student's glucose monitor, considered an emergency exception to the Quiet Zone regulations.

Ruth Bland, the tech director and former school principal in Green Bank, would have loved to make strong WiFi available to the entire school, and she had discussed it with observatory officials. One idea was for the school to install LiFi, which emitted low-range WiFi through lightbulbs, but that was prohibitively expensive. Another idea was to pile an enormous dirt mound behind the school to essentially shield WiFi from radiating toward the telescopes.

"The only thing is, we'd lose our soccer fields," Bland said.

"What's more valuable," I asked, "a soccer field or WiFi?"

She sighed. "Why make me choose? I'd like to have both."

There was another thing.

"I don't know what to do about the electromagnetic sensitive people," Bland said. "They've entered their children into Green Bank School. They came in with this vision that there would be no electromagnetic activity. But you still have hardwired computers, printers, and all sorts of electronics that make the school run. We can't take it out and do papers and pencils."

CHAPTER FOURTEEN

"Behind the Curve"

FOR EVERY ELECTROSENSITIVE who wanted radio quiet, there were probably one hundred residents who wanted WiFi and cell service, and they elected the county's officials. In early 2018, the Pocahontas County Commission passed a resolution in support of cell service throughout the county, a challenge to the very notion of a Quiet Zone. The commission assigned its attorney, Robert Martin, to contact all major telecommunications providers asking them to invest in Pocahontas.

"I'm doing my level best to get another company in here," Martin told me in the spring of 2018. He'd invited me to his house to discuss the new cell service ordinance, and we were swigging Bud Lights at his kitchen table.

"How many cell companies have you written to?" I asked.

"All of them," he said. "I promised the companies that we'll get everybody in the damn county to sign up with them. I'll sign up first! . . . I wrote a letter to everybody and said, 'We have shit for cellphone service here, we want you to come in here, we'll partner with you, we'll help you however we can. Come in here.'"

At our feet were two boxers and a basset hound. In the adjacent mudroom was a 250-pound Vietnamese potbellied pig named Pig, who was snoring. Pig knew how to open the front door and pull a blanket over himself. "I'm the true image of West Virginia, aren't I?" Martin laughed. "I got a pig living in the house." Despite his home literally being a pigsty, Martin was always the best dressed at county meetings, usually wearing tight designer jeans, leather boots, and a crisp dress shirt, top buttons undone and a few chest hairs curling out. A blustery guy, Martin was once jailed in Marlinton for contempt of court for arguing with a circuit judge. He had a history of getting into fights at West Virginia University football games. For years, he'd also operated a hotel in Belize, paying "tens of thousands of dollars in bribes" and putting the payments on his tax returns so the U.S. government could see the corruption he was dealing with (even if he was admitting to violating the Foreign Corrupt Practices Act). Martin came across as a dogged lawyer who knew how to get things done. And he wanted cell service.

"You seen that commercial saying Verizon has more coverage than anyone else?" he asked me. "Pause and look at it real closely, and you'll see right where Pocahontas County is because almost the entire Eastern Seaboard is all yellow [signifying cell coverage] and right there in southeastern West Virginia there's this hunk about this big—it's Poca-fucking-hontas County. I swear to God. Right fucking there we are on Verizon's commercials."

Martin knew well what connectivity was like outside the Quiet Zone. He had earned his law degree from West Virginia University in 1979, married a girl from Marlinton, and started his career in Pocahontas County before becoming a well-heeled insurance defense lawyer in Charleston. He'd gotten his first cellphone in 1986—it was the size of a beer bottle, with a three-foot-long antenna, and it went to bed with him every night. That attachment ended in 2012 when he moved back to Pocahontas, where he only carried an iPhone so he could listen to music in his truck. I asked if he was concerned about the impact of cell service on the electrosensitives.

"Wackos that are afraid of their brains getting fried and all that?" he responded. "Yeah, I know about them."

"They see Green Bank as a haven," I said.

"So? *So?*" He said he wasn't going to let the electrosensitives keep Pocahontas "behind the curve" for cell service.

"But I'm here *because* you're behind the curve," I said. "That makes this place unique."

"You think we want to deal with stone knives and axes for the rest of our existence? You're like these fucking people who move in here and don't want it to change, that it? We have people who have moved here in the last five to ten years and they don't want anything to change. They've 'discovered' Pocahontas County and now *nothing* can change. Well, *fuck,* that ain't the way of the world. We have limitations because of the observatory, because of our topography, because of our insignificant population. But we need to do what we can as government entities to make things available to people."

"Of course," Martin added, the cell service would have to comply with the Quiet Zone.

"We believe in the observatory, we don't want to fuck with them," he said. "Right now, as you and I are sitting here bullshitting, they're up there looking for fucking E.T. And I want to give them every opportunity to do that. But I've got emergency services I've got to render in this county."

In addition to trying to bring in cell service, Martin was assisting the county's emergency services director, Michael O'Brien, to improve communications. The 911 center in Marlinton had difficulty broadcasting any emergency radio communications toward the northern end of the county, where Green Bank was located. O'Brien found a partial solution by installing an internet-controlled radio system just north of Green Bank in the town of Durbin, but it had minimal range and failed altogether when internet or electricity went down. Pocahontas was also one of the only counties in the state unable to adopt a "smart radio system" that integrated radios with smartphones.

On the off chance that someone made an emergency 911 call from one of the county's few pockets of limited cell service, authorities had an especially hard time pinpointing the person's location. "We had a dispatcher spend two and a half hours on the phone one night with a lady that was trapped in her car in a creek," O'Brien told me. "She didn't know where she was or how she got there. We were just keeping her calm while we sent the department to look in all the areas that had cell service."

ACCORDING TO DELOITTE, a 10 percent increase in mobile penetration increases total factor productivity—a key component of economic growth modeling—by 4.2 percentage points over the long run. In Pocahontas, businesspeople like Kenneth "Buster" Varner felt they needed all the help they could get to keep the county's economy puttering along, which meant bringing in cell service.

I first met Varner in early 2017, while eating breakfast at the counter at Station 2. A heavy, jowly man, he had leaned over and asked, "Do you think the gravy is too salty?" As we shoveled down heaping plates of biscuits and sausage gravy, he told me about his various businesses. Aside from owning Station 2, he operated a half dozen enterprises involved in logging, excavation, towing, septic pumping, and auto repair. He was also a fire chief. I told him that

I imagined a lot of headaches trying to manage all those things within the restrictions of the Quiet Zone.

"You have to realize that we never had cellphone service when everybody else had it, so it wasn't anything to us," Varner said. "It'd be more convenient, of course, if it was so you could use your cellphones all the time. But it's a unique place to live where you don't have them, and we take a little pride in that." He noted how the observatory provided jobs and shared its resources, such as lending one of its diesel generators to a funeral home during a recent power outage. "That to me means a lot," Varner said. "And having the largest telescope in the world out your back door, that's a pretty neat conversation piece."

"People can get ahold of me the old-fashioned way," he added. "Call me on the landline or come look for me."

Spending more time with Varner, however, I realized that he was hardly a Luddite. When we met again months later in his cluttered office, I found it hard to keep his attention. He kept glancing down at his iPhone to check texts and alerts he was receiving over WiFi. When he took a call, I was left to stare at a poster of a busty woman in a red bikini and firefighter helmet. When he finally put down the iPhone, I told him I was confused. Hadn't he said he took pride in not using a cellphone?

"I thought it was rude to have a smartphone," Varner said of his "old" perspective, apparently from just a few months earlier. "I do a lot of business on that phone, more than I ever thought in my wildest dreams that I would do." I asked if he could ever go back to living without one. "Wouldn't want to. It's so handy."

Varner had an AT&T data plan. He used Siri. He wished all his employees and volunteer firefighters could always be connected through smartphones. Instead, because of the Quiet Zone, he'd invested more than $30,000 in a specially approved radio repeater system to allow his workers to communicate via low-band radio. "I don't want the observatory to close and for people to lose their jobs," he said, "but it'd be more convenient for everybody."

BY ITS OWN ESTIMATE, the Green Bank Observatory contributed about $30 million to the local economy—a calculation based on the ripple effects of fifty thousand annual visitors, one hundred full-time staff, and forty seasonal workers. But the county's far bigger economic driver was tourism.

Snowshoe Mountain Resort, which boasted the best skiing south of the Mason-Dixon Line, employed 150 full-time employees and hundreds more part-time seasonal employees. A half million skiers descended on Pocahontas every winter, turning the county into Airbnb's biggest market in West Virginia. The ski village could sleep up to ten thousand people a night, more than the population of the entire county, and all those overnighters made up the majority of the county's one million annual tourist visits. Snowshoe contributed about $1 million annually in hotel/motel taxes, whereas the Green Bank Observatory, as a tax-exempt federal property, paid zero taxes. (The observatory's 2,700 acres factored into the federal government's payment in lieu of taxes to the county, which totaled around $850,000 a year for all 310,950 acres of federal land in Pocahontas.)

Tourism funded the county. And tourists wanted cell service and WiFi, according to Cara Rose, director of the county's tourism bureau. She had an intimate understanding of the Quiet Zone, since she'd previously worked for fourteen years at the observatory in marketing. After becoming tourism director in 2011, she created a handout titled "Welcome to the Quiet Zone," which listed places in Pocahontas to find cell service and WiFi. She also turned the tourism office into a workspace where anyone could connect to free WiFi. While Rose personally appreciated the area's radio quiet—and even unplugged her house's WiFi to force her daughters offline—she said tourists felt anxious if disconnected for more than a day or two.

And it wasn't just tourists. Young adults who moved back to Pocahontas County found it challenging to readjust to the quiet. Chelsea Walker, who had taken a job in the tourism department after graduating from West Virginia University in 2016, recalled a camping trip when she was so desperate to get online that she scrunched herself in a rear corner of the family SUV, straining to hold her device in just the right way to get a bar of service. "That's when I had a gut check and said, 'This is ridiculous, I have a problem,'" she recalled. Returning to Pocahontas, she deleted her Tinder account. Beyond the lack of cell service, there were so few people in Pocahontas that she already knew everybody who might appear on Tinder within her geographic radius. Her classmate Makinsey Cochran faced a similar period of reckoning when she returned in 2016. She once insisted on leaving a square dance early so she could check whether her boyfriend had texted her. She worried he'd be mad if she was out of contact for too long.

Friends from college couldn't understand why Cochran and Walker had gone dark. The idea that cell service would be restricted by the presence of a world-renowned astronomy observatory didn't fit traditional stereotypes of the area. In a memoir titled *At Home in the Heart of Appalachia,* which was partly set in Green Bank, author John O'Brien (another of Bob Sheets's brothers-in-law) noted that outsiders often acted like they "knew all about Appalachia—strange, backward people struggling with grinding poverty in a devastated landscape; feral hillbillies; Hatfields and McCoys, black-bearded moonshiners murdering one another. That or quaint folk in strap overalls and granny dresses playing fiddles while they clogged around." Many visitors had no knowledge of the Quiet Zone and simply assumed the area's limited connectivity was because it was a backwoods forest of uneducated hillbillies who simply didn't need cell service or WiFi.

Some visitors canceled their hotel reservations or left in a huff upon realizing they would be without cell service, according to Nelson Hernandez, who operated the Old Clark Inn, the county's longest-running lodging (open since 1924), down the road from the tourism office in Marlinton. Visitors were so habituated to being constantly connected that they were hesitant to explore or bike along the Greenbrier River Trail without cell service. As a solution, Hernandez invested in a GPS device for visitors to borrow. It acted as a kind of emergency beacon, allowing the user to send preset messages like "Help!"

Oscar Martinez, one of about two dozen migrant workers employed at a lumber mill just north of Green Bank, recalled the grim shock upon realizing that he'd landed at the one place in America where cell service was outlawed. "We have cell service everywhere in Puerto Rico," he said. "We came here and we're like, 'What the fuck is that shit?' All the Puerto Ricans come and think they're going to the city, a nice place with discos and parties. But there's nothing anywhere here. There's one bar, and you go and they all look at you like they want to fight with you."

George Nader, another migrant worker, said he used to watch *Supernatural, CSI,* and *House* on his smartphone back in Puerto Rico. "Here, I cannot even watch one episode!" he said. "I started calling back home to ask my friends what happened in the series."

ONE OF THE AREA'S FEW cellphone antennas was in Marlinton, where a low-range AT&T transmitter was hidden in the steeple of

City National Bank. Mayor Sam Felton considered it his responsibility as an elected official to bring in another cell carrier. He was sympathetic toward the electrosensitives, but they represented a minority view.

"Maybe Pocahontas County is not remote enough for their situation," Felton told me when I stopped by his wood-paneled office. "I understand they moved to the county because of the Quiet Zone at Green Bank. Well, we're *already* being sensitive to the requirements of Green Bank, and to do any less is to keep the majority from being serviced because of a very small population."

A round-faced man with a thin mustache, Felton's religious beliefs permeated the mayor's office. A pocket-size copy of Gideons New Testament sat on his desk, along with the book *God Bless America: Prayers & Reflections for Our Country*. His business card had an image of Jesus. On the wall was a framed copy of Luke 16:10: "He that is faithful in that which is least is faithful also in much." He opened town council meetings in prayer. He believed the Bible held "the answer to all our problems." I once heard him start a sermon with a solo rendition of the hymn "Love Lifted Me," which was moving, in a simple and sincere way, though also awkward because I was one of only four people in the church. Felton appreciated the supernatural, the mysteries of life, the unseen powers that govern the world. But that appreciation didn't extend toward the unproven illness of electrosensitivity.

"I don't want to go backward," Felton said. "I want to keep going forward." Forward meant keeping the lights on and bringing in more cell service.

When I got up to leave, Felton motioned to a piece of paper taped by his door. It was a computer-printed photo of clouds.

"Don't you see anything?" Felton asked me.

I was at a loss. "Um, are they cumulus clouds?" I said.

Felton told me to step back. He pointed to a shadowy spot in the cloud. "That looks just like the face of Jesus," he said with utter seriousness. He pointed to another cloud and said, "Now look there, doesn't that look like a woman's face? This really got me thinking of how the Bible says we're surrounded by a cloud of witnesses."

Felton believed he'd photographed the face of Jesus in the clouds. Was it all that much different from another person believing he or she was sensitive to WiFi?

Felton had also asked the attorney Robert Martin to look into the clouds. "I thought he was going to lose his shit when I said, 'Sam, I see a Jack Russell terrier,'" Martin told me.

CHAPTER FIFTEEN

"Mountain Justice"

THE TRAIL FORKED, both directions disappearing into densely thicketed, untrodden woods. Jenna and I had already forded three swiftly moving rivers, each swollen from an overnight downpour that left the ground muddy and our clothes soaked from brushing against waterlogged branches. We were eight miles into the Cranberry Wilderness, the U.S. Forest Service's largest wilderness area east of the Mississippi, and our only means of navigation was a paper map, soggy and disintegrating. Our legs were lacerated and covered in itchy rashes from stinging nettles. Which way would get us out? This was not what Jenna had in mind for her birthday.

We were lost, again, in Pocahontas County, consumed by that feeling of being untethered from society and without connectivity. All around us, a never-ending forest muffled sounds and erased footprints. The area felt simultaneously beautiful and foreboding, welcoming and hostile—a rugged, rolling terrain of secrets. After fifteen miles, Jenna and I finally emerged from the forest, which was perhaps better than others could say.

"There's people who went missing in Cranberry Wilderness that we've never found to this day," Michael O'Brien, the county's emergency services director, told me. He mentioned how a Cessna

414 crashed in the 47,815-acre wilderness in November 1995, going missing for six years. The body of the pilot—who was flying solo—was never found. That potential to disappear had attracted people to Pocahontas over the decades. "You wouldn't believe the number of people who come here to end their lives," O'Brien said. "It's secluded. It's wild. It's beautiful. You can come here and say, 'Nobody's going to find me for a long time, if ever.'"

The quiet of Pocahontas had initially sounded so idyllic. Then I started hearing about this darker side, finding myself engrossed with the stories and drawn to several mysterious, unsolved killings. Pocahontas was an area of extremes, and I felt sucked toward its poles. I hadn't come to the Quiet Zone to investigate decades-old murders, and Jenna reminded me that such was far from my original premise of looking for a place outside the bounds of modern connectivity. But maybe it was all connected, so to speak. With solitude came isolation. With disconnection came the inability to call for help. Perhaps the unsolved killings were a symptom of how justice played out in an ultraquiet place.

TINY HEMPCRETE FIBERS filled my nose, lining my nostrils and clogging my sinuses. All afternoon I'd been climbing up and down a ladder while shoving mushy handfuls of hempcrete—an insulating material made of cannabis fibers—into the walls of Bob Must's new home in Lobelia. He and his wife, Ginger, were moving a mile away to "civilization," a property with grid electricity where they wouldn't have to snowshoe up a half-mile-long driveway every winter.

I'd volunteered to help work on the house. The construction manager was a young man named Clay Condon, whom I'd first met up the road at Yew Mountain Center. The son of a back-to-the-land couple, Condon was starting a hemp farm in Hillsboro, which had some humor to it. The hippies were now openly growing what they were always suspected of smoking. (When the hemp farm was featured on the front page of the *Pocahontas Times*, one reader threatened to cancel his subscription, telling editor Jaynell Graham that he wasn't paying to see "dope on the front page." After Graham explained the benefits of CBD oil, the man kept his subscription.)

As we stuffed handfuls of hempcrete into the walls of the house, Must told me how June 25, 1980, had almost passed by like any other day. He'd been running errands in Lewisburg, arriving home after dark. Parking his car at a turnout a half mile below his cabin, he spotted two people lying in the brush. He approached and saw the bodies were deadly still. Then he noticed the bullet wounds.

Since Must didn't have a landline to call the police, he drove to a friend's house to use the phone. But that friend wasn't home, so he drove to another friend's house to make the call. Police questioned Must at the sheriff's station in Marlinton, and he became the first suspect in the killing of nineteen-year-old Nancy Santomero of New York and twenty-six-year-old Vicki Durian of Iowa, who had been thumbing their way to a hippie festival called the Rainbow Gathering in the Cranberry Wilderness, on the border of Pocahontas and Webster Counties.

First held in Colorado in 1972, the Rainbow Gathering was a monthlong camping event with nudists, hippies, environmentalists, peace activists, and other counterculture groups, all loosely organized by the Rainbow Family of Living Light. West Virginia was the first state east of the Mississippi River to host the event, and for many reasons the locale made perfect sense: the area was remote and sparsely populated, surrounded by state and national forest. On the other hand, the community was insular, conservative, and wary of outsiders, especially hippies who were anti-war, anti-marriage, anti-religion, and pro sex and drugs.

To try to prevent the gathering from happening, a group of Marlinton residents—led by a young, buttoned-up lawyer named Robert Martin who would later become county attorney—filed a motion in federal court to stop the government from issuing a camping permit in the Monongahela National Forest. The judge denied the petition. But Martin believed he won the argument, in the end.

"This is the Bible Belt, for God's sakes," Martin told me, nearly four decades after the fact. "I said, 'Judge, you cannot allow this to happen in Pocahontas County, that's a clash of societies, that's a clash between classes of people that are so different that something bad is going to happen.' And something bad *did* happen."

BOB MUST WAS well covered in alibis, but authorities still believed the killer was local, given how gunshots had been fired in the direction of the Rainbow Gathering's camp. It was also hard to imagine an outsider wandering down a series of narrow, winding, disorienting backroads to Briery Knob, where Must found the bodies. No highways pass through Pocahontas. Getting there invariably required going over a mountain pass. If you were in Pocahontas, you wanted to be there, you had to be there, or you were born there.

During a police investigation that stretched more than a decade, multiple witnesses implicated a local farmer named Jacob Beard, who had been in his early thirties at the time of the killings. In the years that followed, he had been charged with animal cruelty for stabbing his mistress's dog and slicing open her cat—among many macabre subplots that included stories of a third woman having been run through a corn chopper. At a 1993 jury trial, two local men testified to having seen Beard shoot the women. Additional eyewitnesses put Beard in the area on the afternoon of the murders. He was convicted and sentenced to life in prison.

But there was a twist: another man had already confessed to the killings. While in jail in 1984, nine years before Beard's conviction, the racist serial killer Joseph Paul Franklin told a Wisconsin Department of Justice officer that he'd killed two hitchhikers in West Virginia because the women said they engaged in interracial relationships. His confession was barred from the 1993 trial, however, because he had subsequently denied responsibility and then refused to talk further.

In a chilling coincidence, Franklin was connected to the neo-Nazis in Pocahontas, though the murders happened several years before William Pierce purchased his mountain compound. As a teenager, Franklin had once carpooled to a white supremacist gathering with David Duke and Don Black, who would become leaders in the movement and friends with Pierce. Pierce would dedicate his book *Hunter* to Franklin and laud the serial killer for doing "his duty as a white man." In turn, Franklin admired Pierce and listened to his white power radio broadcasts.

A *60 Minutes* investigation into the case helped spur a retrial, this time with Franklin's confession admitted as evidence. Of the two eyewitnesses in the first trial, one now said he'd been coerced by police into making a statement against Beard. After less than three hours of jury deliberation, Beard was acquitted. He received a $2 million settlement from Pocahontas County for wrongful conviction. Franklin was executed by lethal injection in Missouri in 2013 for other crimes, having never faced a judge or jury on the question of whether he killed the Rainbow women, which only deepened the suspicion in Pocahontas that Beard may have gotten off the hook.

Eugene Simmons, the longtime county prosecutor, was one of many people directly involved in the investigation who still suspected Beard was guilty. He personally knew Beard, having once hired him to work on his farm. "He was a good guy until he started drinking," Simmons told me. Former sheriff Jerry Dale, who investigated the case and interviewed Franklin, believed Franklin falsely confessed to increase his rank among hate groups, boost his standing among inmates, and potentially get transferred to a lower-security penitentiary in West Virginia, which had fewer Blacks in its prisons as well as no death penalty. "Trust me, I know who killed those girls, based on the evidence," Dale told me, describing Beard as a Jekyll and Hyde character. "There's no doubt in my mind." The attorney Robert Martin was familiar with the men allegedly involved in the Rainbow murders, having served as defense counsel to a Green Bank woman who hired two of them to knock off her husband—they cut off the husband's head. "They were immoral criminal types who didn't give a shit about nothing," Martin told me. That case was not to be confused with the Green Bank man who decapitated his wife, buried her head in the backyard, and tossed the rest of her body down his well.

OF THE TWO PEOPLE who testified in court to seeing Beard kill the women, I was told that only one was still alive. I found him in the same Hillsboro house where he'd been living in 1980 at the time of the murders. Nearly four decades after the fact, however, Winters Charles "Pee Wee" Walton was even more hazy about what had happened. He recalled being in a van atop Briery Knob on a hot day. Beard and some other men were drinking whiskey or another kind of cheap liquor. He heard gunshots. He noticed bullet holes in their van afterward. As for what happened in between, "I never could get it straight," Walton said, "because I was drinking, too."

During the investigation in the '90s, Walton had been roughed up by law enforcement in an effort to jog his memory, with police punching and slapping him, placing a boot on his neck and threatening to strike his testicles. He was also taken to a hypnotist to aid his recall.

"Did it work?" I asked Walton, as we chatted on his porch one evening. "Were you hypnotized?"

"I don't know if I was or not," he said.

"I guess that's the nature of hypnotism," I said. Finally, I simply asked, "Do you think Jacob Beard did it?"

"I thought he did," Walton said, "but I'm not really sure."

The only other person alive who would know what had happened was Beard himself, so I looked him up and called a couple of wrong numbers before reaching him at his house in Florida. In his seventies, Beard was surprisingly willing to talk about a case that had either ruined his life or allowed him to get away with murder, depending on whom you asked. He still kept in touch with friends from Pocahontas and subscribed to the *Pocahontas Times*, but he said he could never return to the county to live.

"A lot of people up there think I'm guilty," he said.

"What would you say to people who think you did it?" I asked.

"They're going to believe what they believe. At my age, I don't care what they say."

Over the phone, I ran through the evidence that I'd heard against him: that on the second anniversary of the murders, when the case was still cold, he'd called one of the slain women's parents to offer condolences; that Walton had testified to seeing him pull the trigger; that he'd been known as a volatile and heavy drinker; that he'd been charged with mauling his girlfriend's pets. Beard said he drank only the occasional beer (which contradicted local testimony and records of a 2006 arrest for drunken driving), that he didn't kill his ex-girlfriend's pets (even though the animal cruelty case was documented), and that he didn't even know Walton (which seemed unlikely, given that everyone knew everyone in the community).

The only thing he didn't deny was that he had called one of the slain women's parents in 1982.

"I said I was sorry for what happened to their daughter," Beard said. "I just was hoping to get the investigation started again. I had two daughters and I wouldn't want to feel what those parents felt, to be called up in the middle of the night and told their daughter was murdered."

There was an awkward silence.

"It sounds like you believe I'm guilty," he said.

"I honestly don't know what to believe," I said. "But you can imagine where I'm coming from, having spoken to so many people in Pocahontas County and hearing everything from one side so far, and that's why I need to speak with you, because I need to get your side."

"My side is I wasn't there. I didn't do it."

In recent years, two true crime books had also concluded Beard was innocent. Regardless of who did it, *somebody* got away with it.

SWIRLING AROUND THE CASE was the notion of "mountain justice," a phrase that I heard repeatedly. Bob Sheets, who had been on the second grand jury for whether to bring Beard to trial (and called it a "no-brainer" decision to send the case to court), described mountain justice as local people taking matters into their own hands, a pact of silence falling over those involved, and an acceptance that the law played out differently in Appalachia.

As examples of mountain justice, I heard of an old widow who shot the legs of a man spotlighting deer in her orchards, and a young woman who shot her deadbeat husband four times through the front door; neither woman was charged, because mountain justice had already been served. Sheets told me about a particularly infamous case that ended with three people dead, making the front page of the *Charleston Daily Mail* in 1973. A Green Bank woman was having an affair with a man who happened to have a pirate-like hook for a hand. The hooked man got into an altercation with the woman's husband in the high school parking lot after a graduation ceremony, stabbing the husband with a knife and killing him. He was arrested but released on bond, and whispers began circulating that he had a target on his back. Weeks later, the hooked man (who was from a neighboring county) had a rendezvous in Green Bank with the woman. They were ambushed on Wesley Chapel Road and both shot dead, their bodies left in the road along with five shell casings from a twelve-gauge shotgun.

Many locals had an idea of who killed the couple, according to Sheets, but "there was no investigation. Mountain justice."

Eugene Simmons was the prosecuting attorney then, and he told me there was evidence that at least three people were involved in the killings, with multiple vehicles used to block the couple's getaway. It would have been up to him to open a criminal investigation, but he said an overnight rain had simply washed away all the evidence. "Generally speaking, everybody knew who it was, but there was no evidence," Simmons said. "I dropped the case." (His own legal ethics weren't entirely unquestionable, as he'd had his law license suspended in the '90s for fleecing clients.)

Simmons himself had been on the receiving end of mountain justice. In 1972, he was riding a tractor through his hayfield when he heard his engine backfire. When he heard the sound again, his arm and leg started bleeding. He'd been shot. "I jumped off and started running, made it to the fence and passed out," he recalled. "Some people drug me out and brought me to the hospital." He'd nearly been killed by a man named Jack Biggs, who was angry that Simmons was representing Biggs's wife in a divorce proceeding.

"He shot the whole bottom of my arm off," Simmons told me. "If he'd been a quarter of an inch higher, it would have taken the arm off and I would have probably bled to death." Simmons hadn't been armed then, but he said he'd kept a firearm by his side ever since.

"WE HAD A SHIT TON of murders around here back then," the county attorney Robert Martin was telling me. Not that Pocahontas was dangerous, he added. Crime was low. Neighbors looked out for one another. People weren't scared to live there. Andrew Must, who grew up near where his father found the murdered Rainbow women, told me that the killings were rarely discussed—not because they were taboo, but because the incident was so unrepresentative of everyday life.

"Do you know that right now you can walk out there in my driveway and all three of those vehicles have the keys in them?" Martin said. "Do you know we don't lock our houses? There's not a lock on this house that works. Do you think I worry about that? I don't lose a goddamn minute's sleep over it."

"C'mere a minute," he said, getting up from his kitchen table. I followed him down the hallway, up the stairs, and into a spare room. He pointed to a dozen rifles leaning in the corner. None was locked up, which was Martin's whole point. He was unconcerned. He picked out an antique rifle that had been carried by his great-grandfather into battle for the Confederacy.

"And you've got a sword," I said, noticing a steel blade among the gun barrels.

"You ever seen *Braveheart*?" Martin said. "You know at the end of the movie when Brendan Gleeson throws that sword that flies through the air and sticks in the ground? That is *the* sword from the movie."

"How'd you get that?" I asked.

"I bought it," he said matter-of-factly. He pointed to another sword. "You ever seen *The Last Samurai* with Tom Cruise? See this? That's *the* practice samurai that Tom Cruise used in that movie."

Martin had movie memorabilia signed by Robert Redford and John Travolta, walls of expensive artwork including a lithograph signed by Kurt Vonnegut, and shelves of exotic rocks on display. "That's a dinosaur egg," he said, picking up an ovular rock. "See this?" He pointed to another one. "It's a dinosaur turd." Hidden away in a chest, he also had a collection of Nazi memorabilia, including an authentic Nazi armband, a Nazi officer's ring, and souvenirs from the 1936 Olympics in Nazi Germany. He'd purchased it during an estate auction in Charleston. "A bunch of skinheads were after that stuff," Martin said. "I wasn't going to let those sons of bitches have it. I decided I'd buy it and burn it before I let those bastards have it."

Between all the rifles, rocks, and movie mementos was a mounted cover of *West Virginia Executive* magazine from 2009 featuring Martin in a dark suit, with a big smile and shoulder-length blond hair. He'd cut his hair when he returned to conservative Pocahontas County in 2012, traded his suits for jeans, and strapped on a pistol.

"I was the assistant prosecutor around here and I went after everybody," he said. "I was a prick, and I wore a big fucking gun every day while I was in the prosecutor's office. Think I worry about that now? No! There's nothing to fear here. That's endemic to West Virginia. It's just rural America."

THE SUMMER OF 1980, in the days after the Rainbow murders, Simmons and Martin—who then ran a private law firm together in Marlinton—visited the Rainbow Gathering to see what was happening with the "gaily bedecked and unbedecked thousands," as the *Washington Post* described attendees. Dressed in business suits, Martin and Simmons had to wade across a creek to reach the forest encampment, so they removed their shoes and socks and rolled up their pants. Many attendees had removed *all* their clothing, and they were walking with arms outstretched toward Martin and Simmons to place flower wreaths around the lawyers' necks.

"The girls came out and they were all hugging us!" Martin said, chuckling at the memory. "And a guy came out and he starts to hug us. You gotta remember this was the eighties. You didn't even do bro hugs back in the eighties."

Martin recalled going to the infirmary at the Rainbow Gathering and hearing a loud, booming voice. He turned around: it was a six-and-a-half-foot-tall doctor named Hunter "Patch" Adams, years before he would become a celebrity thanks to the movie starring Robin Williams. Dressed as a clown, pockets full of gags, his hair dyed neon colors, Adams was running a medic tent for the hippies. He happened to have purchased 310 acres in Pocahontas County earlier that year with a plan to build a free hospital to serve rural Appalachia. Martin had been Adams's lawyer for the property's $67,000 sale and title transfer.

Nearly four decades later, locals were still waiting for Adams's long-promised hospital to open, despite millions of dollars having been donated to the project.

"It's just a joke," Simmons said of Adams and his Gesundheit! Institute. "They've never treated anybody to my knowledge. Just took the money."

"I know he's raised a ton of money for this hospital," said Sarah Riley of High Rocks, "but the hospital doesn't exist so I'm not sure where the money is going."

"They don't really do anything for us," said Joseph Smith, the former Marlinton mayor. "They didn't do anything when I was mayor, and I was on the town council for twenty years before that and they didn't do anything."

In the quietest place in America, could one get away with murder *and* fraud?

CHAPTER SIXTEEN

"Where's the Hospital?"

MY INITIAL IMPRESSION of Pocahontas County as a quiet paradise was initially deepened by the presence of Patch Adams. What could be more perfect than a clown with a medical degree running a free hospital in an area in need of health care?

I'd seen the 1998 film *Patch Adams* as a boy, watching it on VHS at my grandparents' house. I loved the story of a goofy doctor played by Robin Williams challenging the stuffy medical establishment to have more compassion and humor. Advertised as being "based on a true story," the feel-good movie took in $200 million at the box office and closed with a line about how Adams had purchased land in West Virginia, where "construction of the Gesundheit Hospital is currently underway" and "a waiting list of over 1,000 physicians have offered to leave their current practices and join in Patch's cause."

The movie was loosely based on Adams's 1993 book *Gesundheit!* about his life and mission. An army brat who grew up in Germany, Japan, and the United States, he had earned his medical degree in 1971 from Virginia Commonwealth University and then helped run a free, home-based medical clinic in northern Virginia for a decade. He left that project with the aim of opening a full-scale hospital on land he'd purchased in Pocahontas.

"In a way, finding the property was like reaching the Promised Land," Adams wrote in his book. "It injected a renewed sense of purpose. Our fund-raising efforts, just barely begun, shifted into high gear." He focused on public donations, raising awareness for his project through dozens, if not hundreds, of media interviews, including a 1988 appearance on *The Oprah Winfrey Show* to promote his "forty-bed free hospital in West Virginia." After drawing in tens of thousands of dollars in donations, Adams hired the Vermont-based architect David Sellers to design a "silly hospital," which was to have an eyeball-shaped eye clinic and an ear-shaped examination room, according to his book, whose movie rights were quickly optioned by Universal Studios. The blockbuster film brought Adams international fame. Overnight, he could command $20,000 for speaking engagements. His hospital appeared all but assured of becoming a reality.

Then he started asking for more money. In his book, Adams had said he needed $5 million to build his hospital. By 1999, the amount had ballooned to $50 million—half for construction, half for operating costs—and Adams's vision was coming under criticism from "medical professionals, health-care volunteers, and even his ex-wife, who says he has done a poor job managing donations to his nonprofit Gesundheit! Institute," according to a 1999 article in *People* magazine. Linda Edquist, who divorced Adams after twenty-six years and accused him of cheating on her for five of them, told the *Washington Post* in 1998 that most of the money being raised for Gesundheit! was being spent on Adams's own salary and rent.

Despite the negative press, Adams kept doing his thing, delivering paid presentations, organizing volunteer clowning trips around the world, and raising money for a hospital in West Virginia. Two decades after the movie's release, in his mid-seventies, Adams was still talking about his soon-to-open hospital.

DOWN ANOTHER NARROW, winding road of Pocahontas County, I turned into a dirt driveway with a wooden sign that read "Gesundheit! Institute." An antique fire truck was parked by the entrance—presumably a nod to the comedy bit about clowns fighting fires. I rounded a bend to a large, grassy clearing with a caretaker's house, a three-story workshop building, and a pond that was said to be a popular spot for skinny-dipping. Farther up the dirt road, I drove by two yurts, a cabin, and a multi-domed living facility in the shape of an elaborate, ostentatious Russian dacha, with two onion-shaped towers and two breezeways each in the shape of keyholes. The building looked like an Appalachian version of St. Basil's Cathedral.

I was greeted by a young man named Adam Craten, who gave off the vibe of Shaggy from *Scooby-Doo*. He said he was the property caretaker while Adams was away, which sounded like most of the time. While the Gesundheit! Institute's website referred to "our home in West Virginia," Adams actually lived five hundred miles away, in Illinois. Adams's Twitter page listed his location as "Hillsboro, WV," but he visited Pocahontas only a handful of times a year.

Scattered around the property were gardens of kohlrabi, kale, and collard, peppers, sweet potatoes, and pumpkins. Craten seemed most proud of his 225 feet of tomato plants, which he estimated would yield six hundred pounds of tomatoes. "Patch says the land hasn't been this thriving since the eighties and he sleeps really good at night because we're here and doing all this awesome stuff," he said. "We try to be as sustainable as possible."

Craten showed me inside the skeleton of a new twenty-thousand-square-foot building under construction. Built by Andrew Must and

Clay Condon, the second-generation back-to-the-landers from Lobelia, the building was to be a future library and teaching facility with classrooms, an industrial kitchen, and sleeping quarters for forty people. Adams had already invested $1 million in the building. Progress was on hold while the organization waited for several million dollars more to materialize, Craten said.

He next led me inside the multistoried dacha, which had curving walls, winding stairways, and so many bedrooms and beds that I lost count. All those quarters were intended for people who came to Gesundheit! for paid events or to volunteer. A group of undergraduates from the University of Notre Dame had just visited for a fall break service learning trip, tending the gardens and going clowning at a nursing home in Lewisburg. Students had also come from New York University, the University of Maryland, and the University of Michigan to volunteer, according to Craten.

But I was curious what the students were supposed to be learning. Online speaking agencies described Adams as having *already* "constructed a 40-bed hospital/healing center in rural West Virginia." Gesundheit!'s website, meanwhile, said the nonprofit was still working to complete "our most important project yet: the Gesundheit Hospital," and that "all donations take us one step closer to realizing Patch Adams' vision, offering free healthcare within a community fostered on the importance of happiness, silliness, love, creativity and cooperation to achieve true health." But where was this budding hospital? And where was Adams?

"Some people come in and say, 'Where's the hospital?'" Craten said. He acknowledged that Adams had a unique, almost baffling mission, but such was his prerogative. "Patch raises most of the money for this place, he can do whatever the fuck he wants."

Maybe. But Adams had promised something very different when asking for donations. In his book, he described building "fully equipped, acute-care inpatient and outpatient services that can handle all aspects of rural medicine on a drop-in or scheduled basis. Emergency room, general surgery, X-ray, laboratory, pharmacy, ophthalmology, gynecology, acupuncture, dentistry, physical therapy, and many other specialties will be represented." That was pretty specific—and pretty far from the reality on the ground. Money was clearly flowing into the organization, it wasn't just being spent on any medical facilities, as far as I could tell.

With no sign of Adams, I sent him an email via his website. Days later, my laptop rang with an incoming Google Voice call. It was Adams, calling from his home in Urbana, Illinois.

BY ADAMS'S OWN ADMISSION, he was "almost never there," meaning Pocahontas, because he needed to be near an international airport for all his clowning trips, which he called "humanitarian aid missions." He claimed to have led more than two hundred such trips, including to war zones and refugee camps. He only intended to move to West Virginia once his hospital was complete. Given his age, I asked if he thought he'd live to see that happen.

"I am optimistic," he said. He just needed another $70 million, he added. He was confident new donors would step forward, and he said a $300 million donation appeared imminent. (When I checked back with Adams in early 2021, he was still waiting for that big donation. But on the bright side, he added, he'd been nominated for a Nobel Peace Prize. He didn't appreciate when I told him Donald Trump's son-in-law had also been nominated.)

Adams's mission was still to build a live-in medical facility and "eco-village" that would revolutionize health care by showing how services could be provided at a fraction of the cost of a traditional hospital. Medical and cleaning staff alike would receive a monthly salary of $300, he said. He would also save money by eschewing medical malpractice insurance, which he said introduced mistrust to the physician-patient relationship and stifled doctors from using "creativity" and "intuition." West Virginia is among the minority of U.S. states that do not require doctors to carry malpractice insurance.

Adams had a list of physicians ready to come to Pocahontas, though he conceded that a number of them had died in the time it was taking his project to come to fruition. Still, he wouldn't budge from his goal of opening a full-fledged medical facility that would include an "enchanted playground with interconnected tree houses."

"But wouldn't a simple clinic better serve Pocahontas County?" I asked.

"We're not building it for the people of Pocahontas County," Adams replied. "We're building a hospital to show you can do it at 5 percent of the cost."

To date, Adams had shown he couldn't build a hospital at *any* cost. He called his original $5 million project estimate a forty-year-old guesstimate. Plans had changed. Costs had inflated. "We don't have any idea what the plan is now because we're going to wait until we're funded and then get together and make the plan," he said, in one of many statements that left my head dizzy. I couldn't make sense of his circular argument for needing money to build a hospital whose costs had not been pinned down.

"Why should somebody be convinced to donate now," I asked, "when after four decades there's still no hospital to show for the donations that have come through?"

"*What* donations?!" Adams growled, his cheerful facade crumbling. "Almost all the money donated is money *I made.*"

Adams denied that Universal Pictures ever donated $500,000 to Gesundheit!, as had been widely reported in the '90s. The Hollywood studio "didn't give me shit," he said. Ninety percent of all financial pledges to Gesundheit! failed to materialize, he added, though he didn't have a dollar amount for that other 10 percent. Most donations were under $100. He repeatedly asked me to make a donation, which he said was the least I could do for taking up his time.

I said public tax documents showed more than $15 million going into Gesundheit! from gifts, grants, contributions, and gross receipts from admissions and merchandise at his paid appearances.

"I'm sure I made *at least* $15 million in the years after the movie," Adams said. But that was personal income, he said, quickly adding that he poured much of his money into his organization. He still charged up to $20,000 for speaking appearances.

When I later rechecked his nonprofit's tax filings, I tallied nearly $20 million going into the organization from 1997 to 2018, as well as about $800,000 in annual outgoing expenses, with the net result being that Gesundheit! sometimes filed an annual loss—though net revenues over two decades were still in the millions. Nearly 40 percent of annual expenses went toward administration and fundraising, which was high, according to Charity Navigator. Charities ideally spend no more than 25 percent of their budgets on nonprogramming costs. When I asked Adams to help me understand some larger expense items on his tax returns for compensation, wages, and travel, he told me he didn't pay attention to those numbers. "I've never looked at our tax filing," he claimed. When I asked if he could put me in touch with his accountant or another person who might explain the numbers, he flatly said no.

Adams said he needed another $2 million in the short term to finish building the teaching center that I'd seen under construction. Once opened, it could host classes that might be a source of revenue.

But if that building alone cost several million dollars, I asked, why didn't he just invest the money in an actual hospital?

"How dense are you, Stephen?" Adams said through what sounded like gritted teeth. "All of it is *one thing*! The buildings, the house, the staff aren't different than the hospital. Can you get that in your noggin? If you don't have a residence for the staff then you can't have a staff and so you don't have a hospital even though it's built! It's a *complete project*!"

I was confused about another thing. The back of Adams's book said the Gesundheit! Institute "has treated more than 15,000 people for free" in West Virginia, which sounded to me like a reference to his hospital project in Pocahontas. Adams said that was a typo. From 1971 to 1983, he and a few friends had treated fifteen thousand people in northern Virginia and the eastern panhandle of West Virginia out of home-based medical practices. But why wasn't the book cover clearer? I asked.

"I don't *give a shit* what's on the back of a book!" Adams snapped back.

Or maybe the book was never corrected because it wasn't in Adams's interest for the truth to be known that he'd never treated a single person in Pocahontas. There is no hospital. There never was a hospital. There likely never will be a hospital, according to people who know Adams.

MAUREEN MYLANDER WAS a struggling freelance writer when Adams asked her to ghostwrite his book in the early '90s. She told him that she only did coauthorship and equal split of the earnings. He agreed. She researched and wrote the book over two years as a side project to her full-time job. Several evenings a week she went to Adams's house in Arlington, Virginia, and interviewed him for a few hours.

The output of their collaboration helped create the mythical image of Adams as an idealistic and beneficent clown doctor building a hospital in Appalachia. According to the publisher, Healing Arts Press in Vermont, the book has sold more than 250,000 copies, most of them coming after the success of the film *Patch Adams.*

"It was not the best book I ever wrote, but it was the most successful, and it's because somebody wanted to make a movie out of it," Mylander told me.

Some things changed between the book's 1993 and 1998 editions, however. As people became disenchanted with Adams and dropped out of Gesundheit!, their names were removed from the book. While Mylander's own name stayed on the cover, even she grew suspect of Adams.

"He just never struck me as a builder of hospitals when I didn't even see a clinic down there in the early nineties," Mylander said. "Anybody could have bought some bandages and had somebody there with some nursing care."

Mylander recalled being at Adams's fiftieth birthday party in the mid-'90s, watching as friends presented him with a $10,000 check toward the Gesundheit! Institute. She overheard Adams say, "I wish there were a few more zeroes on it."

"That gave me the idea, 'What amount would be enough?'" she said.

At one point, Adams invited the local doctor Bob Must to join Gesundheit!'s board of advisers alongside the celebrity doctor Benjamin Spock and the peace activist Norman Cousins, who promoted laugh therapy. The invitation held some bitter humor. In the years after Adams purchased his property, Must went to medical school and became an osteopathic doctor; practiced medicine around the region at hospitals, prisons, and clinics; opened a private practice in Hillsboro; led the board of Pocahontas Memorial Hospital; *and* was named the 2012 Outstanding Rural Health Provider of the Year in West Virginia—all while Adams failed to treat a single person in the county. Must told me it had cost him about $71,000 to open his clinic, which saw thousands of people over its decade in operation—including, ironically, people from the Gesundheit! Institute.

Must declined to join Gesundheit!'s board. Looking into the organization's bylaws, he determined it was essentially a dictatorship under Adams. He suggested to Adams that, instead of trying to raise tens of millions of dollars for a full-service hospital, he might instead focus on offering wellness promotion and disease prevention at a clinic that could start seeing people immediately. But Adams would not be swayed from his vision.

Danette (Brandy) Condon was once part of Adams's organization. A petite woman from Detroit, she had helped scout out the property in 1980 and then lived on the land for several seasons. In 1981, she moved to Lobelia, raising two sons with a back-to-the-lander. In the years since leaving Gesundheit!, Condon had studied midwifery, become a member of the Midwives Alliance of West Virginia, and assisted in hundreds of home births, doing real medical work that Adams had failed to provide in Pocahontas.

"I think it's a big scam," said Allen Johnson, the former libraries director. As a social worker and administrator at a nursing home in Marlinton from 1987 to 1993, and then a mental health social worker from 1994 to 2000, Johnson was well aware of the county's health needs—and of Adams's empty record in servicing them. "For thirty years they've been talking about this free hospital," Johnson said. "Where is the hospital?"

Gesundheit! volunteers occasionally went clowning at the Marlinton nursing home and participated in local activities. One time, they offered to repaint the lookout tower at Droop Mountain Battlefield State Park, so superintendent Mike Smith prepared the materials. "After they painted the easy logs at the bottom and off the stairs, they decided, 'Nah it's awful hot,' and they went down to the river to go swimming," Smith said. "Me and my one worker had to finish the rest of the tower." I saw it as an analogy for what had been happening with Gesundheit! for decades: good intentions and meager results, with the community left to do the real work. "As for as actual real results of people coming in and being healed and helped, it just hasn't happened," Smith said.

In the 1990s, Sheriff Jerry Dale started hearing complaints about Gesundheit! and opened an informal fact-finding investigation. "Patch was a celebrity after the movie," Dale told me. "My opinion was that he was using the notoriety of the movie to bring people in and donate money and start a hospital and this and that. What I was concerned about was people that were disillusioned by the whole reason that they came here to begin with." But the investigation led nowhere. "I kind of walked away from that whole thing feeling sorry for Patch and feeling sorry for a lot of the people that were there," Dale said. "It seemed that they were lost, looking for a place in life and something meaningful."

Dale had more pressing issues. With only one sheriff and a handful of deputies tasked with responding to accidents, giving speeding tickets, making drug busts, collecting taxes, conducting welfare checks, serving court papers, providing courtroom security, monitoring a neo-Nazi group, *and* solving murder cases in a county that was nearly the size of Rhode Island, law enforcement never stood a chance.

THERE'S A SCENE in the '90s television series *Twin Peaks* in which FBI agent Dale Cooper, who is enchanted by the quaintness of a sleepy mountain town—the cherry pie, the Douglas firs, the "way of living I thought had vanished from the earth"—learns of a dark undercurrent.

"There's a sort of evil out there," the sheriff tells Cooper. "Something very, very strange in these old woods. Call it what you want. A darkness, a presence. It takes many forms but . . . it's been out there for as long as anyone can remember." Later, Cooper encounters the darkness himself, and there's no going back. There's no unknowing.

That's what discovering the many layers of the Quiet Zone felt like. The vision that drew me in turned out to be a mirage. The Quietest Town in America was full of WiFi and smartphones. The astronomy observatory was partly a cover for a government spy facility. The electrosensitives seemed to be fleeing something in their lives aside from electromagnetic radiation. The free hospital for rural Americans who desperately needed access to health care was a joke.

At this point, I was not surprised to hear there had once been something of a sex cult in Pocahontas. Of course there had been. And of course it had been advertised as an idyllic commune.

The Zendik Farm, about twenty miles south of Green Bank, appeared on the surface to have been a kind of artists' collective where young people tilled the land, ate organic food, and published quirky literature and music that they sold on city streets and at festivals nationwide, which earned them a bit of publicity. On MTV in 2004, the pop singer Christina Aguilera wore a Zendik T-shirt with the slogan "Stop Bitching, Start a Revolution."

On the inside, the commune was a dystopian nightmare for some. A former member named Elliott Kelly described the Zendiks as a hippie version of Nxivm, the upscale self-help firm based in

New York that was uncovered in 2017 to be a cult where women were manipulated into becoming the leader's sex slaves. That's how women were treated by the Zendik Farm's founder, Wulf Zendik, according to Kelly. Another former member named Helen Zuman has also questioned whether Wulf Zendik was "scheming, from the beginning, to gain sexual access to nearly every post-pubescent female" in the commune.

"I knew that underneath it there was lots of dark stuff going on," Zuman told me. "I didn't think it was wrong. I just thought, 'This is what has to happen if you want to start a revolution.'"

Wulf Zendik had been born as Larry Wulfing, first establishing his anti-capitalist commune in California in 1969. At times numbering up to seventy people, the group migrated around the country—to Texas, Florida, North Carolina, and finally, in 2004, to West Virginia. The group was attracted to Pocahontas County's low population, ecological health, and forests—in essence, its quietude—according to Zuman, who was then a member and did research to help find their property outside Marlinton.

By the time the group arrived to the Quiet Zone, Wulf Zendik had died and his widow, Arol Wulf (born Carol Merson), had assumed control. Through authoritarian fear and Big Brother–esque oversight, she oversaw all the group's decisions—be they sexual, financial, or otherwise. Zuman had just graduated from Harvard with a $13,500 grant to explore alternative communities. She handed over most of the money to the group. She was encouraged to have multiple sex partners, forbidden from having monogamous relationships, and required to seek permission for many aspects of her sexual life. A likely factor in her contracting herpes during that time was that condoms were forbidden in the commune; the ban was lifted in 2004, though members still had to get permission from the group's central authority to use one. Zendiks also didn't believe in HIV or AIDS. Illness and death were caused by having bad "vibes."

But from the outside, the Zendiks just looked like another group of hippies.

"Class group, I enjoyed those people," said the county prosecutor Eugene Simmons, who was also a farmer and sold the Zendiks alfalfa hay. "Those people were a hippie type but not rowdy people. They participated in the community, did flowers and stuff, never had any problems with them."

For years, Arol Wulf wrote a monthly gardening column for the *Pocahontas Times* and tended flower boxes along the main street through Marlinton. Before she died in 2012, she asked a local woman to keep up the flower beds, and they still lined the sidewalks when I rolled through town, although by then the group had dispersed. Their property, known as Wulfsong Ranch, went up for sale in 2013 for nearly $1 million. Rumor had it that John Travolta looked into buying it but passed because it couldn't support an airstrip.

Instead, the ranch was leased out to a family looking for a place that was "cheap, remote, and inaccessible," as they told me. When I visited, I saw some of the eclectic items that the Zendiks had left behind: colorful informational pamphlets, homemade musical instruments, and a six-foot-tall stone monument by a pond where Wulf Zendik's body was still buried.

In a shed, I found a huge, ornately framed canvas painting of an astronaut standing on the moon and looking mournfully toward Earth. The label read "Post-Ecollapse Apology." I asked if I might keep the painting—a memento of the fringe ideas that fermented in the Quiet Zone. The family was happy to see it go. I later learned the name of the ex-Zendik who'd painted the canvas and emailed him to ask if he wanted it back. He said no. He'd since become a professional dog portraitist in New Jersey.

CHAPTER SEVENTEEN

"The True Epitome of Darkness"

WEAVING AROUND POTHOLES and downed branches, my car jerked up and down and scraped over rocks. It was late spring of 2018, and the dirt road to the National Alliance's mountain compound was even more washed out and rutted than the previous fall, when David Pringle had told me he was planning an alt-right festival for Hitler's birthday. April 20 had come and gone, however, and I'd heard nothing about any gathering.

I drove through the entry gate, which was propped wide open. A Toyota Corolla and a beat-up Subaru were parked outside the cottage where Pringle and his wife had been living.

"Hello?" I called into the house. "Hello?"

A grinning face appeared in the window.

"C'mon in!" said Jay Hess, whom I'd met before. "I'm just finishing painting this bathroom."

I poked my head around the corner. Holding a wet brush, Hess proudly pointed to a fresh coat of white paint on the bathroom trim. He said he was turning the cottage into a "VIP headquarters," "like a bed and breakfast or something."

"Wow," I said. "You must have a lot of people living here?"

Actually, Hess was alone. Pringle had departed in early 2018 after a series of incidents, namely the four-wheeler accident and his hosting of "unsanctioned" gatherings, to the irritation of the group's chairman in Tennessee. Pringle had relocated to Nebraska to work in a gunsmith shop, bringing his long-range rifle with him.

The number of people on the property had dropped by two-thirds—to one—but Hess nevertheless insisted the Alliance was on the verge of a turnaround, recruiting "people from around the country" to live there in exchange for providing labor to rehab the decaying buildings. "We expect them to come any day now," he said, trying to sound optimistic. "They'll be fixing up the infrastructure, all the normal things people do to keep everything looking nice."

A dryer in the pantry buzzed. Hess made a show of taking out the laundry and folding it neatly, as if it were just another beautiful day in the neighborhood. I'd heard things weren't so tranquil. He had been recently busted for shoplifting a pair of pants and shoes from the Men's Shop in Marlinton. (Hess said it was all a misunderstanding.) He seemed desperate for cash, not to mention company. After living in Florida for thirty years, he'd had a tough winter in Appalachia, with the snow piling up so deep as to strand him at the compound for weeks. He'd done a lot of reading in the cold.

As for the National Alliance's plans for a massive alt-right gathering, Hess said the agenda was pared down to a "little festive birthday celebration" for Hitler. A dozen people attended. "We got together and had a nice meal inside the main building," Hess said. In recognition of Hitler's vegetarianism, he added, "the women prepared a wonderful eggplant Parmesan."

HESS FINISHED PAINTING the trim and called it quits. As we walked out of the cottage, a short, wiry man in dirty jeans tromped down the wooded hillside with a burlap sack slung over his shoulder. It was Bryan Dewitt "De" Thompson, husband of the woman with the two Rottweilers who'd led me up to the compound a year earlier. The Thompsons lived in the RV a half mile down the road.

Thompson tossed the sack in his Subaru. Hess and I walked over and looked inside. It was full of foraged fiddleheads and ramps.

"Do you have a secret spot where you know to get ramps?" I asked.

"Don't need to be that secret because people are lazy, they won't get out and do this stuff much," Thompson said. He lit a cigarette.

Only in his fifties, Thompson looked ancient. Glasses rested at the tip of his nose. His hair was white, with a yellow tint from chain-smoking. He had no teeth—he lost most of them from doing meth, and the rest fell out during chemotherapy for squamous cell carcinoma. He took a swig from a container of vodka. A bowie knife was slung on his belt.

Though Thompson foraged on National Alliance property, he said he was not a member. His late father, Boyd Thompson, had owned the surrounding land. (In fact, we stood just off Boyd Thompson Road.) He said he'd lived in the area long enough to have met William Pierce, whom he described as someone who "knew how to live here" because "he didn't get into nobody's business." One time, however, Thompson did almost shoot Pierce's German shepherd when the dog charged him. Thompson had taken out his gun, had lifted the safety, and was about to fire when Pierce called the dog off.

Thompson said he'd also met the other local celebrity, Patch Adams, and gone skinny-dipping "with a bunch of beautiful fucking women" at the Gesundheit! Institute.

"What's this?" Hess chimed in.

"Patch Adams," Thompson said. "The movie."

"Robin Williams? Really? Geez."

"You live too protected of a life," Thompson said. He added that he knew many of the back-to-the-landers and hippies associated with Adams, and he'd even attended the 1980 Rainbow Gathering, where he'd been greeted by naked women and gotten "a screaming case of it" from their food.

Oh, and he knew who committed the unsolved Rainbow murders.

"It was a couple ignorant rednecks and they picked up a couple of hippie girls, said, 'Hey, y'all want to get high?'" he said. "Jacob Beard, he was the motherfucker who pulled the pistol . . . There was like one, two, three, four people I know *for a fact* was there because at one time or the other they got drunk and *blah blah blah*. You listen to everybody's stories and then you can pretty well decipher what happened."

"Did you ever hear Beard admit to the killings?" I asked.

"Hell no, he's not that stupid," Thompson said. "Even if he would have said it, it would have been so far back and in a time before cellphones and recording devices. Anything he's ever done I'm sure he'll take to the grave."

"If Beard killed those girls with so many people watching, how could he convince them all to keep quiet?" I asked.

"They were terrified," Thompson said. "He'd have sold his fucking farm and paid somebody to kill you. Hell, my dad was going to pay somebody to kill *me* at one time."

Thompson said he also knew all about another double killing in Lobelia that I'd been hearing about. It was a bizarre case from 1975 involving one of the first hippies to arrive in the area: a young man named Peter Hauer from Pennsylvania, who was a respected caver and secretary-treasurer of the American Spelean History Association. In 1970, Hauer had purchased a Lobelia farmhouse that had a large cave in the backyard. Thompson remembered Hauer as "intelligently ignorant"—smart about what he knew but socially awkward, with "an exaggerated sense of purpose in saving mother nature and all."

Soon, strange things began happening that signaled Hauer was unwelcome. He found a snake in his mailbox. Then sticks were shoved down his horse's throat. Then the horse was killed, its head bashed in. His goats' ears and bellies were slashed open, their innards spilling out. Laurie Cameron, another of the early hippies, had some farm experience and helped sew up one of Hauer's goats that "was gashed open along its flank." A veterinarian's daughter told me she recalled her father being called to Hauer's house late one night to find "limbs, pieces of trees, everything down the horse's throat."

Hauer came to believe his antagonist was a local man named Tommy McNeill, who worked on a nearby farm owned by Boyd Thompson. A police investigation led to the arrest of McNeill, who in early 1975 was sentenced to six months in Weston State Mental Hospital. Shortly before McNeill would have been released, two young men went missing in Lobelia. One was Hauer. The other was his friend.

On June 4, 1975, a West Virginia University honors student named Walter Smith had been riding his ten-speed Schwinn bicycle home from Watoga State Park, where he had a summer job. He never arrived. Smith was the son of a National Steel Corporation executive, so his disappearance triggered a major search. Nothing turned up.

Hauer, who had seen Smith the day he disappeared, told his ex-girlfriend over the phone that "he was also fearful for his life," according to Henry Rauch, an emeritus geology professor at West

Virginia University who knew Hauer and investigated the case for a 2018 article in the *Journal of Spelean History*. Hauer disappeared on June 9, the day after the phone call.

After no sign of Hauer for three days, police entered his house and found a typewritten letter in which the author confessed to having murdered Smith and stashed his body in the cave behind the house. According to the letter, Smith had been "in the wrong place at the wrong time." The letter said Hauer's own body would "eventually be found in a cave in the nearby hills."

That evening, police uncovered Smith's partially buried body in the cave. He had been shot in the head above each eyebrow and once in the neck with a .25-caliber pistol. Smith had also been sodomized, probably prior to his death, according to the West Virginia University medical autopsy report.

Police found bloodstains on a pair of Hauer's clothes in the house. A caver found .25 cartridge shell casings underneath his porch; Hauer was known to have purchased a .25-caliber pistol after the attacks on his animals, according to Professor Rauch, whose research informed much of what I knew about the case. Years later, children would discover Hauer's pistol under a rock in a nearby stream. Law enforcement concluded Hauer raped and killed Smith.

But where was Hauer? Cavers scoured the area. The FBI issued a fugitive from justice warrant and conducted a national search. Authorities also sniffed around the farm of Boyd Thompson, who'd been known to have an open dislike of Hauer because he was an "arrogant" outsider who outspokenly opposed hunting and other aspects of local culture, according to De Thompson. (Boyd Thompson died in 2015.) Law enforcement even dredged the Thompsons' pond to look for evidence—namely, Smith's missing bicycle. They found nothing. County prosecutor Eugene Simmons invited spiritual mediums to aid the hunt. Two psychics performed a séance and pointed on a map to where Hauer's body would be found. As Simmons told me, "They came pretty close."

Nearly six months later, on Thanksgiving Day 1975, a boy named Larkin Dean was hunting with his father when he came across a pair of boots on the ground. He looked up: high in a tree was the head and upper torso of a body hanging from a noose. The corpse had been dangling for so long that it had decomposed and split in half, with everything from the waist down falling to the ground.

The state medical examiner identified it as Hauer's body. Simmons concluded that Smith and Hauer had gotten into a fight over a woman, that Hauer had shot Smith, and that Hauer then hanged himself out of guilt. He speculated drugs were also involved. But this didn't explain why Smith was sodomized. As for the cruelty directed toward Hauer's animals, Simmons said that was just because Hauer was an outsider. "They've got a little animosity in rural areas against outsiders," Simmons told me.

De Thompson had another version of the story: it was no murder-suicide. Hauer had been framed for Smith's killing and then hanged, he said. "Maybe Peter Hauer pissed off the wrong motherfucker and the other people just happened to be in the way, if I was going to take a stab in the dark."

"But why would Hauer leave a suicide note saying he'd killed Smith?" I asked.

"I'd say Peter Hauer would have probably confessed to fucking Sister Mary or Mother Teresa by the time whoever was talking to him got done. He pissed off the wrong person."

Over the years, a rumor had emerged that Smith's missing bike was hidden inside the chimney foundation of the Thompsons' old home, but De said that was a rumor he'd started himself on the school bus.

"My dad's lots of things, but stupid ain't fucking one of them," he said. "Keeping a piece of evidence would not have ever happened."

His phrasing pricked my ears: he wasn't denying that his father was involved, only that his father wouldn't have kept any evidence. So, Boyd *was* involved?

Thompson laughed. He said he and his family had been out of state when Smith and Hauer disappeared. "All I can tell you is we went to Florida," he said.

I was incredulous. A random ginseng hunter in the Appalachian woods held the secrets to at least four of the county's mysterious murders?

"I got a pretty good idea, let's just put it that way," Thompson said. Then he shrugged, because people went missing all the time in those days. "This is karst country," he added, referring to the region's topography of caves and sinkholes. "People disappear." There were more unsolved killings than anybody knew about, he alleged, but they weren't investigated because "you've got to have some good family and somebody to push for you or they don't give a fuck."

I didn't know if I should doubt Thompson because he was a drug-dealing ex-convict, or if his shady history made his testimony all the more believable. Sensing that I wanted to hear more, Thompson offered to take me into the cave where the young man's defiled body was discovered in 1975. Hess quickly said he wanted to join. With more than two thousand feet of passageways, the cave would take several hours to navigate through, Thompson warned. He'd been inside before and the interior chambers were "sort of squirrelly."

He told me to return the next day in proper caving attire: work pants, a helmet, and two flashlights. "Because you don't ever go in a cave," he said, "with only one fucking flashlight."

THOMPSON AND HESS were standing outside when I pulled up in the morning. Five hundred miles away, Jenna was less than thrilled about my caving plans. "I'm a little nervous for you," she'd emailed. I was a little nervous, too. Thompson's criminal record included arrests for grand larceny, dealing drugs, and illegal possession of a Walther P22 semiautomatic pistol that he'd pointed at his wife. Was he up to something with me?

Thompson hopped in the passenger seat of my car; Hess sat in back and leaned forward excitedly. Linda followed behind in her Subaru with their two Rottweilers. As we rumbled down Boyd Thompson Road, I said I'd just come from the high school, where I'd told a class that I was going caving in Lobelia. The students had sounded scared for me. "People *die* down there!" one girl had said. Thompson laughed. In telling the story, I was trying to hint that people knew where I was and that they'd be looking for me if I went missing.

As we drove through Hillsboro, Thompson asked to stop at McCoy's Market. He went inside while I waited in the car with Hess. I slouched down, tucking my head in like a turtle, anxious that somebody might see me driving around with a neo-Nazi and a guy who (by his own count) had been arrested seventy-five times, as well as shot twice. Thompson got back in the car with a bag of fig bars and water bottles—sustenance for our expedition.

We turned right onto Lobelia Road. Houses gave way to fields with Texas longhorn cattle. Pasture turned to forest as the road steepened and wound around Caesar Mountain, passing by the house of Larkin Dean, who had stumbled upon Hauer's body in 1975. (It so happened that Dean's brother-in-law was Jacob Beard, the man initially convicted of the Rainbow murders.)

De Thompson told me to park by a bridge. We were at the site of Peter Hauer's old farmhouse, which had years earlier burned down. We strapped on headlamps. Hess held a flashlight. Thompson also wore his bowie knife and carried a backpack with the water and fig bars. He led us across the road, past the foundation of Hauer's old house, over a small brook, and up a hillside to an opening in the ground about the size of a basement bulkhead. It was known formally as Lobelia Saltpeter Cave. A weathered sign warned that the cave was "Closed to All Access." Thompson squinted at the words and laughed about how signs didn't tell him what to do. He entered first. I followed. Hess went last. The darkness swallowed us.

I WOULD LATER circle back on my own to Larkin Dean's house, finding him smoking a cigarette outside. I wanted to hear what he remembered of finding Hauer's body, and whether he thought Hauer hanged himself.

"Why would Peter Hauer string himself up here when he lived over there?" Dean replied in a tone that made me feel like a moron for even raising the question. Calling it a suicide was a convenient way for authorities to write it off, he said.

Many local people saw evidence that Hauer was framed. In Hauer's farmhouse, police had found the suicide letter beside a bag of groceries, which seemed like an odd purchase for someone planning to die. And why would Hauer hide Smith's body if he was going to confess to the murder anyway? Why would Hauer say his own body would be found in a cave, only to hang himself in the woods? Professor Rauch noted that Hauer's autopsy determined his right leg was broken, and while it was unclear if it had been broken before or after the hanging, it would have been impossible for him to climb a tree and hang himself with a broken leg.

Theories swirled. Hauer's story became somewhat legendary in caving circles, earning the lead-off spot in a book called *True Tales of Terror in the Caves of the World,* which suggested that Hauer and Smith had been involved in a Wiccan cult in Lobelia and their deaths were connected to witchcraft or satanism. Another theory was that McNeill had gotten out of prison and exacted his revenge. Underscoring how strongly the caving community believed in Hauer's innocence, the National Speleological Society in 1979 established the Peter M. Hauer Spelean History Award, which is still given every year. Awards typically aren't given in memory of murderers.

Standing outside Dean's house, three dogs sniffed at my feet. He warned me not to pet the small one with the crazy eye. Dean recommended that I talk to a neighbor, Bill Wimer, who lived farther up Caesar Mountain, second trailer on the right. Wimer owned the property where Hauer's body had been found.

I found a tall pile of corn husks in Wimer's yard. I knocked on the trailer door, and a voice yelled for me to come in. Opening the door, I was hit with a wet blast of aromatic steam. A woman poked her head around the corner. "Well, come in!" she said as if she knew me. Her name was Debbie Wimer. Her father, Bill Wimer, was hunched over a chopping board at the kitchen table, cutting the kernels off what seemed like hundreds of ears of corn. He wore a hat that read "It's Hard to Be Humble when You're from West Virginia." Debbie was boiling jars at the stove in preparation for canning corn.

I said I'd been told they might help me better understand the strange case of Peter Hauer. Debbie's jaw dropped. She said she'd *just* been thinking about Hauer. Days earlier, she had been foraging for ginseng when, out of the blue, the song "Dang Me" got into her head. She'd started humming:

Dang me, dang me
They oughta take a rope and hang me
High from the highest tree

"I started singing it out loud and then I got to thinking, 'Why is that song coming in my head?'" She later realized she'd been standing near where Hauer's body had been found. "It still creeps me out," she said. She pointed to her bare arm. "I got chill bumps!"

There was a paranormal undertone to Debbie's story, a note of mysticism that highlighted how people here lived alongside the spirits of the dead. She had been seven years old when Hauer disappeared. Her father knew more about the investigation. He spoke in mumbles, so Debbie acted as his translator, and there was a lot of back-and-forth over which side of Bruffey Creek Hauer's body had been found.

"I know who'll know about this," Debbie said. "Aunt Betty!" She dialed her aunt in North Carolina and thrust the phone to my ear.

As I stood in the Wimers' trailer amid the thick steam of boiled corn, Aunt Betty told me that, based on what she knew from when she lived in Pocahontas, Boyd Thompson had likely killed Smith and then framed Hauer. Boyd Thompson, of course, was the father of my cave guide.

THE TEMPERATURE FELL 20 degrees as the cave opened to a space about fifty feet wide and twenty feet high. Chunks of ice were frozen on the ground, even though it was 60 degrees outside. We could see only as far as our headlamp beams.

"This is so cool!" Hess said, his voice echoing. "A wild cave! My first one!"

"There's a hell of a breeze blowing through here," Thompson said.

"Tells you there's an opening somewhere," I said.

We marveled at a rainbow-colored stalactite forming on the smooth ceiling. The ground was covered in cobblestone-size rocks amid ten-foot-deep crevices. I scanned for signs of where Walter Smith's body might have been uncovered in 1975.

"I keep thinking I'm going to find a bone or something," I said.

"Anything's possible," Thompson cracked.

We walked farther, one hundred feet, two hundred feet, three hundred feet . . . My mind was left to re-create the four-decades-old police search for Smith's body. My headlamp shone onto a dirt patch, which I imagined as the place where police might have first uncovered the feet, then the legs, then the naked body of Smith.

"Why would anyone bury a body back here?" I asked.

"You mean, why even bother if there's a note?" Thompson said. "Turns you 180 degrees, don't it?"

The cave narrowed. The floor and ceiling sloped toward each other. Thompson began crawling. I followed on my hands and knees, grunting as I slid forward, stopping at a sudden drop-off. My headlight shone straight down twenty feet.

"There's a set of steps over here," Thompson called out. His light cast on a stone stairway leading down the subterranean cliff.

"I don't know if I feel like going much further," Hess called to us. I looked back and saw him wedged in a tight section. "I think I've gone as far as I want to go." His light faded. He was gone.

Thompson and I were alone. I now understood why he'd recommended bringing an extra flashlight, and I wished that I'd taken his advice. If my headlamp died, I'd quickly be disoriented and unable to feel my way out of the cave. We weren't sure if we should go left or right at the bottom of the cliff—both ways were simply blackness.

"Hey, turn your light off for a minute," Thompson said.

He switched his headlamp off with a click. I reluctantly did the same. *Click.* The absence of light was almost breathtaking.

"This is the true epitome of darkness," Thompson said.

In nearly a decade of living without a cellphone, in my years of visiting the Quiet Zone, I had never experienced such a void of sound, light, and life. I'd finally found absolute quiet. And it was terrifying. *Here's where De kills me,* I thought.

"It seeps inside of you," I said.

"You can almost feel it," he said.

It was dark as death, silent as eternity, as if I'd fallen into a sensory deprivation tank. Time slowed in the absence of basic perceptions. In those endless seconds, I imagined Thompson unsheathing his knife, winding back his arm, and preparing to murder me as Hauer and Smith had been murdered four decades earlier. Trapped in my mind, cut off from my senses, I was overwhelmed by irrational fear. I crouched to my knees, thinking this might help me avoid Thompson's first swing.

Click. His headlamp shined into the abyss.

"I just wanted you to see what dark was," Thompson said, and he stepped farther into the cave.

WE CAME TO a fifteen-foot-deep crevasse, scurrying down and climbing up the other side on an old wooden ladder that was missing some of its rungs. This led to a circular room with a pile of charred wood that had been hauled inside the cave, presumably to keep somebody warm. Water dripped from the ceiling. The tunnel rose steeply to a muddy area of dirt and weeds. We'd reached a dead end. Thompson seemed lost. It'd been at least thirty years since he was last in the cave.

"I'd like to make it through this son of a bitch," he said, lighting a cigarette. He opened his backpack to pull out two water bottles, passing me one. "We're going to have to try another hole."

We circled back to the stone stairway. Beyond it, the tunnel narrowed, then opened into a squarish room the size of an office cubicle. Carvings on the wall had dates going back to the 1930s. Thompson spotted a garbage-chute-size opening in the back. He got on his belly and began slithering through it headfirst. "You son of a bitch," he grunted. I pushed his boot to help him forward. His bowie knife got snagged, and he had to tuck it close to his body to squeeze through. His boots disappeared.

"Watch your fucking head!" he called back to me.

I followed, wedging myself into the five-foot-long tunnel, which seemed barely large enough to roll a watermelon through. The glorified worm hole opened up to a crawl space with a long row of stalagmites that looked like dinosaur teeth. I couldn't see Thompson anywhere.

"Are you sure this goes somewhere?" I called out.

"No," his voice echoed back from another opening.

I inchwormed across the dirt, squeezed through another crevice, and emerged into a tubular room filled with stalactites and stalagmites, some as long as three feet. The weird features looked like arms, reaching out to grab us. Thompson was slouched with his back against the wall. My jacket and pants were caked in dirt and mud.

"This one stalactite looks like a dick, man!" Thompson said, pointing. "A dick with genital warts!"

There was a pile of nuts on the ground, perhaps from a spelunking squirrel. We agreed that Hess wouldn't have fit through the last two tunnels. Thompson lit another cigarette and crawled to the back of the cavity, where the ground rose steeply to another small opening. Soon all I could see were the muddy soles of his boots, the rest of his body having disappeared into another part of the cave system. He came back down. He was still smoking his cigarette.

"I think there's a hole on the backside of this, but it's a lot of belly crawling," he said. His head scraped against the ceiling and a piece of rock crumbled down. Our situation felt precarious, as if our movement through these long undisturbed tunnels could cause the rocks to shift just enough to trap or pin us.

"One time I helped pull a spelunker out of a cave," Thompson said. "He went in six-foot-fucking-four but he was close to eight-foot-tall after a rock slid up against him."

"He died?"

"Well, yeah."

Thompson looked apprehensively at the next hole we'd have to squeeze through.

"Air's getting harder to breathe back in here," he said.

WE RETRACED OUR WAY back to the cave entrance. The light outside appeared ethereal. The forest felt humid and dank, full of life and color. We'd been gone two hours. It felt longer.

Hess and Linda stood by the cars. "If there was an opportunity for De to kill me, it was back in the cave," I said to them, joking about how unnerving it was when Hess disappeared and then Thompson told me to turn off my light to "experience" darkness. Linda laughed; she'd been so creeped out by the cave that she'd refused to even step inside. While we were gone, she'd foraged a bag of ramps and other roots. Thompson grabbed one of the plants and held it to my nose. "Sarsaparilla," he said, and dropped it in my hands.

Back at the Thompsons' RV that afternoon, I took out my wallet and offered De a five-dollar bill, which was all the cash I had on me. "For the water and fig bars," I said. "And for your time. I wish I had more to give you."

Thompson cocked his head sideways and frowned. "How about you take it and get some good greasy city food and think about me," he said. "I can't have it too spicy!"

Chickens clucked around our feet.

"Well, thanks for the water, the food," I said.

"Aw fuck, you're more than welcome."

Thompson turned and walked across his yard—past the baby-blue trailer where his chickens roosted and into the rickety RV where he lived with Linda and their Rottweilers. I felt guilty for having been ashamed to be seen with him and concerned he might rob or kill me. He had only been generous. The guy had nothing. He didn't even have teeth. All he had were stories. And yet he didn't want anything from me—just wanted to share in my enthusiasm for the weird history of his county. I drove down the long dirt driveway and onto the road home to Green Bank. The passenger seat was brown with cave dust.

PART THREE

QUIET END?

> "Be quiet!" said Jesus sternly.
>
> —MARK 1:25

CHAPTER EIGHTEEN

"A Do-or-Die Situation"

JENNA AND I TURNED INTO the Sheetses' mile-long driveway, crossing over a creek and passing alongside a horse pasture. Bob and Elaine waved from the porch of their white farmhouse. They'd invited us to join them for Thanksgiving of 2018, and we'd eagerly accepted.

We wanted to be proper guests and bring a dish, but our options were limited after several days of camping at the New River Gorge. At the last minute, we'd swung by the IGA market in Marlinton and purchased two sad-looking, machine-pressed pies in plastic clamshells. Elaine politely accepted our offerings, though we got the sense that prefab pies were not quite worthy of the Sheetses' table. Our wannabe dessert was tactfully disappeared, never to be seen again that evening.

We followed Bob toward the kitchen, where their three adult children and extended family were chatting. After introductions, Bob asked if we wanted to see the deer that his youngest son, Jed, had shot two miles deep in the woods that morning. Of course we did.

In the barn, an eight-point buck lay atop a tractor, its glassy eyes staring into space, its rear haunches severed and hanging from a rafter, dripping blood. The fall 2018 hunting season was in full swing. Earlier in the day at the IGA, we'd parked next to a Jeep Wrangler that appeared to have two broomsticks protruding from the back window. I looked closer and realized they were deer legs. At Marlinton Motor Inn where we were staying, a sign said "Welcome Hunters" and the receptionist was dressed in full camo, as if she'd come directly from the woods. The sidewalk outside our room had a large dark stain, as if someone had gutted a deer there.

Beyond the Sheetses' barn, I could see the Green Bank Telescope standing like a sentinel, its red beacon blinking even brighter into the longest nights of the year. It was surrounded by a patchwork of copper-colored hayfields and frozen pastures, a haven for deer. Hunters had complained of losing access to this 2,700-acre federal property when the government established the observatory. As a way of appeasing hunters as well as culling the deer population, the observatory began hosting an annual hunt in 1993. So many people showed interest that a lottery determined who could participate.

I had attended the previous year's weekend-long hunt, standing by as 172 people armed with rifles and bows prowled the fields and woods around the telescopes. Everyone had been asked to put their smartphones on airplane mode. In a parking lot, I had found a bowhunter named Ryan Repp stuffing forty pounds of still-warm meat into a large plastic bag. Usually he took a photo of his kill and texted it to his wife, but this time she'd have to wait until he got out of the Quiet Zone. "I like it because there's no distractions," Repp had said of hunting in Green Bank. "You don't have to worry about getting a call or email from work every twenty minutes." Behind another building, I had found a father skinning a deer with his two sons. They took out a battery-powered Sawzall and began dismembering the animal. "You want these legs?" the dad asked me, holding up the deer's severed limbs. "You can make a gun rack out of them."

With the observatory still under review by the National Science Foundation, the threat loomed of the telescopes being decommissioned and the area returning to a haven for hunters. Such had nearly happened once, thirty years before our Thanksgiving with the Sheetses. In November 1988, the observatory's biggest telescope collapsed, threatening to bring the entire facility down with it—with lessons for the battle that the observatory faced today in staying open.

THE NIGHT OF NOVEMBER 15, 1988. Elaine Sheets's mother was sleeping in a guest room during a visit to Green Bank when she was stirred awake by a swooshing sound. She got out of bed, walked down the creaky steps, sat in the living room, and stared blankly into the distance, wondering whether a UFO had landed in the backyard. The following morning, she was comforted to know that she hadn't been visited by extraterrestrials. "I heard it collapse! I heard it!" she told Bob and Elaine when they broke the news of the collapsed telescope. "I didn't say anything because you would have thought I was crazy."

Or maybe it *was* a visit from aliens? The front page of the national tabloid *Weekly World News* reported on the telescope's collapse with the headline "Zapped! . . . by Hostile Space Aliens!" Another tabloid, *Sun,* reported:

> One of the world's greatest telescopes was crumpled like a toy by a blast of energy from outer space. The reason? Angry space aliens were fed up with our scientists spying on them! Ufologist Nathan Garvade says there's no other explanation for the mysterious collapse of the 300-foot radiotelescope that turned the powerful instrument into a twisted and mangled mess in Green Bank, WV.

An official analysis determined the scope collapsed because of a fractured steel gusset plate located in a critical spot of the support structure. If anything, it was impressive the telescope operated as long as it did. Designed and built "quick and dirty" (in the words of a former official) between 1961 and 1962 for a bargain $850,000 as part of a rush to expand the observatory's operations, the telescope had exceeded its projected life span five times over. More than one thousand scientists had used it over its lifetime to make some of the most important astronomical findings of the century, including the detection of a pulsar in the Crab Nebula (the first pulsar associated with a supernova). Data from the telescope had also allowed staff scientist J. Richard Fisher to help develop a now famous correlation—known as the Tully-Fisher relation—between the luminosity of a galaxy and its rotation rate, which further established a link between a galaxy's dark matter and visible matter; it was a major contribution to understanding the size and age of the universe. For many years, the three-hundred-foot scope had been the world's largest steerable radio telescope. In 1988 alone, 120 astronomers from around the world had used it.

The scope's demise threw the future of the entire observatory into question. (It also caused alarm at Sugar Grove, where a 150-foot antenna was based on a similar design.) The National Science Foundation had already considered diverting money from Green Bank to fund "higher priority facilities," putting the observatory on warning that its days were likely numbered. Without the three-hundred-foot telescope's capabilities, astronomy in Green Bank was even more limited to an aging collection of outdated instruments.

"If the 300-foot is not replaced, the damage will resonate through astronomy for decades and perhaps centuries to come," the radio astronomer Gerrit Verschuur wrote in *Astronomy* magazine soon after the collapse. He continued:

> The telescope was located in the National Radio Quiet Zone, unique in the world, and if radio astronomers back away from exploitation of this national resource they will lose credibility in the face of other interests who hunger for unrestricted access to critical bands in the electromagnetic spectrum which currently allow us to study the distant universe. Once this Quiet Zone is lost we may never see another created anywhere on this planet.

In retrospect, the telescope could not have collapsed at a better time. By chance, West Virginia was just then assuming a new level of political clout in Washington, D.C. Robert C. Byrd had recently become majority leader of the U.S. Senate and was about to be named chair of the Senate Appropriations Committee, giving him control of the government's purse strings. The state's junior senator was Jay Rockefeller, who held a high-ranking post on the Senate committee that authorized funds for the NSF. Byrd and Rockefeller wielded outsized influence on Capitol Hill for their home state, and they acted fast to help the observatory.

"Green Bank is a unique research site—and an ideal location for a radio telescope—because it is a national radio quiet zone," Byrd said in a statement days after the collapse. "We cannot afford to lose any time in moving forward with replacing this important scientific resource."

Rockefeller toured the site and vowed in a handwritten memo to the observatory's director, "I will do everything I can to help put a first-class dish back in place."

In early December 1988, Jay Lockman and Ken Kellermann were among fifty-six experts on radio astronomy who gathered in Green Bank to decide what replacement might be built, at what cost, and by when. "The National Radio Quiet Zone is Green Bank's greatest advantage," the astronomers wrote in their concluding report. "Considering sites other than Green Bank is pointless." The scientists coalesced around the idea of building the world's largest fully steerable telescope, with a dish 330 feet wide by 360 feet tall—moderately bigger than the old telescope but with a far greater range of motion for scanning the skies. For Green Bank astronomers, it was like having a jalopy replaced with a Ferrari.

The NSF, seeing that Byrd was pulling together money for a big investment in astronomy, wanted to replace the old telescope with a laser interferometer to detect gravitational waves from stellar objects, a project known as LIGO. Byrd had to choose: LIGO or a telescope?

According to a February 1989 memo from the U.S. Library of Congress's Science Policy Research Division, LIGO would require few people to operate, while a new telescope would preserve Green Bank's hundred-plus staff. LIGO's potential detection of gravitational waves would likely result in a Nobel Prize (as would happen in 2017), but that prize would need to be shared among several sites and universities, diluting the prestige to Green Bank. By contrast, a new telescope was "likely to result in many important if not fundamental scientific findings." Scientists expected the telescope to discover new stellar radio sources and extragalactic chemicals, and reveal the nature of pulsars and unknown galaxies "hidden" by dust and light. The only comparable telescope in the world was in West Germany (the country was divided at the time), but the surrounding area was so noisy that German astronomers had been traveling to West Virginia to collect data, according to the memo.

"Green Bank is the only observatory in a truly radio quiet area and there is unlikely to be another such zone," the memo stated. "Pressure exists from radio and TV broadcast interests, and from mobile radio telephone interests to reduce or eliminate the Green Bank zone. Such pressure is likely to increase if the 300 foot telescope is not replaced in a timely manner." In other words, if something didn't quickly fill the void left by the old telescope's collapse, noise would. "I feel the observatory is here facing a do-or-die situation," Green Bank's site director, George Seielstad, wrote in a memo to Byrd's office.

Byrd and Rockefeller threw their support behind a new telescope, which they said would be "the best promise for jobs, education, tourism, and scientific prestige." Byrd called for an emergency appropriation of $75 million. He still had to convince Senator Barbara Mikulski of Maryland, who chaired a subcommittee that oversaw the NSF.

In an April 5, 1989, letter to Mikulski, Byrd touted the proposed telescope's "location in a radio quiet zone which assures a very low level of interference when making astronomical observations. The Green Bank observatory is unique among radio telescopes in the world in this aspect." Additional letters to Mikulski came from scientists across the country, including Bernard Burke, an astrophysicist at MIT who had participated in the site selection of Green Bank in the 1950s and later served as president of the American Astronomical Society. Burke said it would "be virtually impossible" to recreate the Quiet Zone.

Mikulski approved the funding request. But if the astronomy community appeared united around the idea of building a new telescope in Green Bank, such was also a political necessity. It was understood that Byrd would fund a new telescope *only* in West Virginia. In truth, a number of American astronomers believed that a big scientific investment would have been more useful elsewhere.

"There's no question [a hundred-meter telescope] would have been better in New Mexico," Kellermann, the former assistant site director in Green Bank, told me. A high-elevation desert climate was generally more conducive to radio astronomy, which was why the Very Large Array was established in New Mexico in the 1970s. While the Quiet Zone provided regulatory protection from human interference, it could do nothing to change Green Bank's climate of snow, rain, and clouds that inhibited measurement of short wavelengths. West Virginia is one of the cloudiest states in the country, with the city of Elkins near Green Bank seeing cloud cover nearly 90 percent of the year. "But Senator Byrd wasn't going to pay for a telescope in New Mexico," Kellermann said.

Construction of the new telescope took a decade—double the projected time—and went $29 million over budget, putting the final price tag close to $100 million. In August 2000, Byrd cut the ribbon for the Robert C. Byrd Green Bank Telescope. Two years later, he appropriated another $8 million for a new dormitory and visitor facility named after his wife, Erma. If articles about the telescope didn't mention Byrd's name, his assistant was known to call the observatory's staff to, in essence, remind them whom they owed their jobs to.

"The three-hundred-foot telescope's collapse was such a shock that it allowed things to happen that otherwise would not have happened," said Lockman, who kept a giant bolt from the wrecked telescope on his office windowsill as a reminder of how a single weak point could threaten the downfall of an entire facility. It was also a symbol of how what appears ruinous could be lifesaving. "The whole site would probably be closed by now otherwise," Kellermann said.

FAST-FORWARD TO THE PRESENT DAY, with the observatory again facing the threat of closure. With the death of Byrd, Green Bank had lost its great patron, and radio noise was flooding the community, eroding the Quiet Zone that had always made the location so valuable.

With time running out for the observatory to make its case to stay open, private citizens were stepping up to offer support. A West Virginia student crafted homemade dolls that looked like astronomers; they were sold at the observatory's gift shop, with proceeds going to the facility. The observatory started a private membership club, and the first to sign up was William Mullin of New Jersey, who told me he personally donated nearly $1,000 to the facility because he felt connected to the astronomical work happening in Green Bank. He'd grown up near where Karl Jansky first discovered stellar radio waves in 1933, and it was heartbreaking for him to imagine Green Bank's "world-class instrument not being used to significantly advance radio astronomy."

Scientists from NASA's Jet Propulsion Laboratory, Stanford University, Harvard University, the European Southern Observatory, and the California Institute of Technology had written a joint letter in support of continued NSF funding for the Green Bank Telescope, calling it "a unique scientific resource for the US community" with unmatched capabilities. Hundreds of community members had voiced support for the observatory during public meetings hosted by the NSF. Electrosensitives were also speaking up.

"There is a relatively large community of people who have been injured by wireless radiation and who have sought refuge in the Green Bank area," Sue Howard, the electrosensitive from New York, wrote in a letter to the NSF. "Where exactly do you think these people should go if the Green Bank Observatory were to close? For us, this is not about losing a job or having to move. This is about our very survival."

Behind the scenes, Green Bank had a much more powerful ally

than scientists or electrosensitives: the National Security Agency, the intelligence-gathering arm of the U.S. Defense Department.

"Remember, we are only the face and not the power behind [the Quiet Zone]," Karen O'Neil, the director of the observatory, had said during a 2017 conference in Green Bank. "The power behind the National Radio Quiet Zone is increasing the frequencies in which it is interested. For those of you who don't know, the Sugar Grove facility . . . is now owned by the NSA and they are increasing their usage of it as time goes on. The NRQZ is not in any danger."

But that didn't mean the observatory's own financial support was assured.

IN THE SHEETSES' DINING ROOM, two tables were pulled together to accommodate sixteen of us and our heaping plates of turkey, sweet potatoes, mashed potatoes, spinach salad, cranberry sauce . . . There was certainly no store-bought anything. Bob's brother-in-law said a prayer: "Father, Son, and Holy Ghost, whoever eats the fastest eats the most."

After dinner and most of the guests excused themselves, I mentioned to Bob that I'd swung by the *Pocahontas Times* and spoken with editor Jaynell Graham. She had caught me up on local happenings, including a disturbing incident: a woman with electromagnetic hypersensitivity was recently found dead in the woods near the observatory. Graham didn't know the woman's name, only that she was an outsider.

It felt a bit morose to mention such a thing at Thanksgiving, but I couldn't not think about it. Bob immediately knew what I was talking about. He said the rumor was that the woman's body had lain in the woods for days, within sight of passing cars, before anyone noticed.

Could she be someone I'd met? I'd spoken with so many electrosensitives over the two years that I'd been visiting, and all their tales of suffering had clouded together. I'd grown deeply skeptical of their ailment, to the point that I'd stopped being concerned for them. Now someone had died.

CHAPTER NINETEEN

"You've Got to Experience It"

"OH MY GOD!" SUE HOWARD SAID. "So much has happened since I last spoke to you."

I was at a Christmas Day gathering of the electrosensitives. Sue's husband, John, was in the kitchen chopping potatoes for our potluck dinner. He'd arrived that morning, having driven through the night from Westchester—and this time he didn't plan to go back to New York. They would now both be full-time residents of Green Bank. That fall they'd purchased a 255-acre property, which included 122 acres of cow pasture and rolling hills that could protect them from neighborhood WiFi. It was a quiet zone within the Quiet Zone. In a sign they'd made the right decision, their U-Haul had by chance featured a giant image of the Green Bank Telescope—part of the moving equipment company's efforts to promote travel across America. To the Howards, it was providence. They still had to build a house on their new land, so for the time being they continued to rent Diane Schou's mobile home.

"We'll have to take you to our property and walk you all around it," Sue said. "It's a big, beautiful forest."

About two dozen electrosensitives and their spouses crowded inside an old farmhouse attached to an inn called Mountain Quest.

Steaming platters of food crowded onto the tables. We all gathered in a circle. Allan Clark, who'd told me a year earlier that he could make lightbulbs glow with his bare hands, said a prayer. We each grabbed a plate and dispersed into two dining rooms. I sat beside a woman named Clover, with whom I'd once locked arms at a square dance. Across the table was a family of four—the only family of color in the group. I introduced myself, and the mother said her name was Sittul Monna. She was from Bangladesh and her husband was from Haiti. She looked at me more closely and exclaimed, "I've seen you jogging in Prospect Park!"

I had to ask her to repeat herself. *She'd seen me in Brooklyn?*

"I go to the farmers' market every Saturday and I know that I've seen you!" she said.

We were practically neighbors in New York City. And we'd both ended up in the Quietest Town in America for Christmas. It was the family's fourth visit to Green Bank since 2016, after initially coming on behalf of their eleven-year-old son, who they believed had developed electrosensitivity from a vaccine. The boy was recovering thanks to the family's visits to Green Bank, Monna said. They wouldn't be allowing their daughter to be vaccinated, she added.

The family was looking into purchasing property in Green Bank because of their concerns about all the wireless technology in the city. "That kind of environment can kill you," Monna said. In their Brooklyn apartment, she had replaced all her lights with special red bulbs. She never looked at a computer screen without wearing blue-light-blocking glasses. She always wrapped her head in a protective scarf when she rode the subway.

"I can't handle big cities," Clover chimed in. "I fall apart. I can't think straight. Even going into Lewisburg, I start cross-wiring, I start doing the opposite of what I want to do. It's very frustrating." She added that "places with a lot of industry, WiFi," had high rates of cancer and other illnesses—"it's very straightforward."

The others nodded along. Having grown up in an evangelical church and witnessed how worshippers fed off one another's religious zeal, I could recognize the group fervor that happened whenever the sensitives gathered. They helped each other believe in the reality of their illness and encouraged one another to recognize "evidence" of the harm from radio waves. In Green Bank, they had found not just quiet but companionship and validation.

AFTER DINNER, I introduced myself to the inn's owner, Susan Alexander Bennet, who went by Alex. A gray-haired woman with intense green eyes, wearing a frilly green blouse and Santa Claus earrings, Bennet seemed to get on well with the sensitives. She'd turned off the WiFi for the gathering, and as a general policy she encouraged all guests to stay off their devices because "we're really trying to help people break away."

Bennet and her husband, David, had purchased the 450-acre property in 2001, expanding the original farmhouse into a conference center with a two-story, twenty-six-thousand-volume library. They'd moved here from Washington, D.C., where Bennet had worked for the U.S. Department of the Navy as chief knowledge officer and deputy chief information officer for enterprise integration. She held a doctorate in "human systems engineering," which I gathered to be a mix of business management, psychology, and information technology. She was also a Reiki master, meaning she believed she could heal people through energy vibrations. She told me the Quiet Zone was an asset to her work, calling Pocahontas "one of the purest areas" because of the abundance of nature and the lack of cell service. I asked if she could tell me more about how life here was unique. She told me to follow her across the hallway.

In another room, the walls were covered with dozens of framed photos of ghostly black-and-white images. At a table in the corner, Bennet picked up a photograph and showed it to me: it was her dog, Sash, who had died years earlier. Bennet held up another photo: it showed swirling fog. Bennet put the photos alongside each other. The fog had an uncanny resemblance to Sash. She believed the spiritual presence of Sash was captured in the photograph.

"These are not ghosts," she said. "These are messages. These are energies."

Bennet gestured around the room. She had nearly fifty thousand such photos of "energy spirits" or "mysts," as she called them. They appeared on clear, fogless nights when she walked outside, sang into the darkness, and asked for these paranormal energies to manifest themselves. After several flashes of the camera, the orbs would appear. "Dimensional holes open up and thousands of these orbs come streaming out," she said. She'd self-published a short book about the phenomena called *The Journey into the Myst.* It was available on Amazon, she added.

Another hundred people had been able to get their own photographs of the mysts, Bennet said. On the wall was a framed photo of a foggy swirl that resembled Mother Teresa, meaning that Bennet believed the saint's actual spirit had visited Pocahontas County. Why here? Because it was the Quiet Zone, where airwaves were free from interference and people free from distractions, and therefore able to tune in to their surroundings, Bennet said. She showed me another myst that, with some imagination, looked like a person playing the violin. She believed it was the presence of her first husband, a violinist, who had died years earlier.

"Can I play devil's advocate?" I asked. "You know how when you look at the clouds and see different objects?"

Bennet smiled. "Isn't that wonderful?" she said. "Our reality, everything that you do, you create from your own head and from your set of experiences . . . What you see is going to always be unique to you based on your experiences, your beliefs, your values, your mental models, et cetera."

"So how do you know if it's actually your late husband in the myst or if you just wanted to see that?"

"I wouldn't have ever wanted to see him—I divorced him!" Bennet shot back. Then she laughed. "Actually, it was very comforting to see that picture. I thought, 'That's great that he's still nearby and hopefully not holding any negative energies.' But anyway, you can see the face. You can see the violin. It's pretty easy."

While Bennet thought she had discovered an interdimensional portal in her yard, each time I visited Green Bank was like crossing into another dimension, or at least an alternate reality.

When I later asked Bert Schou what he thought of the mysts, he sounded surprisingly skeptical.

"I can't discount it, but I can't totally believe it," he said.

"There's a parallel between you doubting the mysts and other people doubting electrosensitivity, no?" I said.

"Yeah," Bert said. "You've got to experience it."

Before leaving Mountain Quest, I was tasked with vacuuming the dining rooms. Sue Howard said she and the other sensitives were pained by the touch of the machine, so I agreed to expose myself to the supposedly harmful electromagnetic radiation. I also had to roll my eyes. If it wasn't good for them, why was it okay for me? We had all sat under the inn's lights all afternoon, near a lit-up Christmas tree, and I'd even spotted several smartphones at the tables. Yet they couldn't touch a vacuum?

I found Bennet cleaning up the kitchen, and she offered me a glass of water that had been run through a special purifier made of crystals.

"This is with crystals," she said, handing me the glass. I sipped. "This is without crystals," she said, handing me another glass.

I said I couldn't taste the difference.

"Take another swallow," Bennet said, as if she were willing me to taste the power of the crystals.

We were all putting on our jackets when I finally broached the subject of the electrosensitive who had died. There were nods and murmurs. Everybody had heard about the suicide, but nobody knew the woman's name or any concrete details about her death. Kathryn Stauffer, the sensitive from Illinois, said she'd heard the woman was from Virginia and had stayed overnight at the Boyer Motel. Stauffer herself once tried to stay there but was so bothered by its air conditioners that she'd instead slept in her car.

"She probably came here desperate," Stauffer surmised. "She probably got here and thought, 'It's no good here, either,' and didn't take the time to find one of us."

A COUPLE DAYS AFTER CHRISTMAS, I visited the county clerk's office in Marlinton. As I flipped through a huge book with hundreds of pages of death certificates, a name caught my attention: Marianne Roberts, age sixty-five, of Crozet, Virginia. Date pronounced dead: September 14, 2018. Cause of death: "Perforating gunshot wound of the head." How injury occurred: "Shot herself." Place of injury: "Wooded area," five miles from the Green Bank Observatory.

I left the clerk's office, walked down the carpeted hallway, and turned the corner toward the county prosecutor's office. Eugene Simmons waved me inside. He had handled the case—yet another mysterious death over the decades he'd been in the office. He confirmed that Roberts was the sensitive who had killed herself. She was last seen alive on September 11 at the Boyer Motel. She'd checked out, driven a mile up the road to Boyer Hill Mennonite Church, walked down an embankment into the woods, and shot herself in the head. Three days later, her body was discovered by a road crew. Simmons had inspected the scene, finding it apparent that Roberts died by suicide. Police discovered a pistol by her side and several suicide notes in her car. To get more details about the woman, Simmons suggested I speak with the Boyer Motel's owners, Frank and Jane Murphy, the last people known to have seen Roberts alive.

Later that week, I found Frank at his motel, and he told me he remembered Roberts. She'd taken an interest in his cats. "Come up to the auto shop, my wife can tell you more," he said. His wife, Jane, worked at Murphy's Auto, run by their son. Frank added that his son had actually spotted Roberts's body in the woods but hadn't called the police because he'd thought it was a mannequin discarded over the embankment—which sounded odd, at the very least. I didn't know of a single store in the area with mannequins, meaning there wasn't exactly a mannequin glut in Pocahontas. And in any case, why would someone go through the trouble of dragging a mannequin into the woods?

I followed Frank a mile up the road to the auto shop. Jane sat at one of three metal desks, eyeing me warily. Her son, Mike, a stern-looking man with a dark mustache, glared at me from his desk. His wife, Karen, called Frank into a back room. Nobody seemed excited to talk to me.

"I have no idea what happened," Jane said, lips pursed. "I only talked to her a couple times. She came to play with my cats and talk about her cats." Roberts had arrived without a reservation. She used the Murphys' home phone to call her employer in Virginia, letting them know she'd be back in a few days. She left nothing behind when she checked out the evening of September 11.

Karen called Jane into the back room. Then Karen reemerged alone. "You need to leave now," she said to me.

I asked what the problem was.

"I think you're an investigator," she said.

I repeated that I was a journalist.

"We're working here, buddy," Mike growled in a way that made clear I was *not* his buddy. "We're trying to finish a Friday and I've got a pile of shit going on. Just get the hell on out of here."

CHAPTER TWENTY

"The Only Way Out"

ANY SUICIDE IS UNSETTLING. Details around this incident left me feeling especially uneasy, from the fact that Marianne Roberts had driven into the Quiet Zone with the seeming intent to kill herself, to how her body had lain unnoticed for three days—save for the guy who mistook her for a mannequin. Had she killed herself out of desperation? Was she trying to send a message? If so, what was that message?

The death hadn't appeared in the *Pocahontas Times,* but I did find Roberts's obituary online, and it listed the name of a sister in Texas. I looked her up, found a phone number, and called. Leslie Stanga picked up. She seemed eager to talk, as she was still trying to make sense of her sister's death.

"She barely told me she was feeling so bad," Stanga said, "so I was totally blindsided when we found out that she had committed suicide."

The sisters hadn't seen each other in years, but they'd kept in touch by phone, and Roberts had said she felt pain from WiFi and cellphones. Stanga was unaware that Roberts believed WiFi was *literally* killing her.

Stanga helped put me in touch with a handful of Roberts's family and friends, and in speaking with them a picture emerged of an intelligent and ambitious woman, obsessive and intense, idealistic and empathetic, who became reclusive in her final years. Born and raised in Louisiana, Roberts had attended Vassar College in the 1970s and then gone to Harvard University for her MBA, earning her master's degree from the world's most prestigious business school not long after it started admitting women. She worked as an investment banker for the Wall Street firm Dillon, Read. She briefly married and divorced, never having children. In 1986, she abruptly switched careers and became a caretaker at Innisfree Village, a rural community for adults with disabilities near Shenandoah National Park. She got involved with animal rights advocacy in nearby Charlottesville through a group called Voices for Animals, which rescued feral cats and provided free spaying and neutering. Charlee Pawlina, who co-led Voices for Animals with Roberts, described her as "a brilliant woman filled with compassion for everybody" and "one of the most nonviolent people I know." What changed to make her commit such a violent act?

Around 2015, Roberts's health became an issue after she was exposed to a large amount of mold while helping a friend named Donna Middleton move out of an old barn house. "The mold issue led to a downward spiral of things she was allergic to," Middleton said. The same year, Roberts discovered she had benign brain tumors called meningiomas. One side effect is tinnitus, or ringing in the ears, which Roberts later claimed to suffer from. In 2016, Pawlina's husband developed a brain tumor near where he'd always worn a Bluetooth earpiece, which led Roberts to research how the devices emit electromagnetic frequencies. Then she personally started to feel sensitive, becoming disoriented, dizzy, and nauseated when near electronics, with the sensation of her body vibrating, buzzing, and burning.

Doctors could offer Roberts no explanation for her symptoms, so she began researching alternative healing methods. Already a vegan, she cut more foods from her diet, losing about seventy pounds between 2015 and 2018. To water bottles, she applied special stick-on labels that imbued the liquid with "healing frequencies." She consumed a cocktail of herbal supplements, including molecular hydrogen and magnesium.

Roberts spent around $10,000 on protective EMF-blocking gear: hat, gloves, face mask, and specially designed wire-mesh clothing that looked like chain mail armor. Family would later discover hundreds of dollars in unpaid bills for the gear, and Stanga would angrily call one of the companies to accuse it of preying on her sister's paranoia.

Roberts became wary of going outside for fear of being exposed to cell signals. She curtailed her computer usage and began turning off her power breaker. She developed neck pain for which surgery was recommended, worsening whatever suffering she already felt. Monika Kohler, a coworker, recalled Roberts saying she felt enough pain to jump off a cliff. Kohler asked if she had suicidal thoughts. Roberts said no.

She tried going out west to detox. She spent time in Arizona and visited a clinic where she paid thousands of dollars to undergo a weeklong body detoxification. Nothing seemed to help. Then Roberts visited the National Radio Quiet Zone. When she returned, she told friends she was considering moving to Green Bank.

In September 2018, several months after Pawlina's husband died from his years-long battle with brain cancer, Roberts made plans for a trip to the Quiet Zone. She left behind meticulous instructions on how to care for her cats. "It somewhat seemed she was leaving to not come back," said Kohler. Friends and family later learned that Roberts had tried to euthanize her sixteen-year-old cat. When the vet refused, Roberts had responded, "I could die at any moment and I want to make sure this cat is not going to be stuck in a shelter." The vet agreed to adopt the cat.

On Tuesday, September 11, Roberts called Kohler from the Boyer Motel near Green Bank and said she'd be back at Innisfree by the end of the week. When Roberts didn't return by Friday, Kohler became concerned. She called the motel and spoke to Jane Murphy, who said that Roberts had checked out days earlier. A few hours later, Kohler received a call from Murphy saying police had found Roberts's body.

LESLIE STANGA and her husband met with Eugene Simmons, who had collected Roberts's belongings. He gave them the keys to her Prius, the five suicide notes discovered in her car, and the .38 Smith & Wesson revolver found by her side. The Stangas asked Simmons to throw away the gun. Roberts had purchased it just a week before her death from an outdoors supply shop in Virginia. Her family and friends believed she'd never shot a gun before.

"The very idea of Marianne buying a gun and shooting herself—if I was told to name ten things in the world that would never happen, that would be one of them," Charlee Pawlina told me.

When making such predictions, Pawlina spoke with confidence. She was a professional psychic and member of the Minnesota Society for Parapsychological Research who worked under the name Lee Channing with an "energy interpretation" business called Spirits Evolving. She normally charged around two dollars a minute for consultations.

"I've done some famous people," said Pawlina, who was in her seventies. "But I am not going to invade Tom Cruise's privacy!"

She said she could tap into people's energies to read their future and had worked with law enforcement to solve murders. In one case, Pawlina divined the initials of the killer. A year later, according to Pawlina, a man with the same initials pleaded guilty. She also worked with dying animals and ailing people to help bring them peace and resolution. And she spoke to dead people. After learning of Roberts's death, in fact, Pawlina had "tuned in to" Roberts's last moments alive.

"She was calm and clear the moment she pulled the trigger," Pawlina said. "I did not feel any fear. I felt resolution in her mind."

Roberts had become religious in her final years, hanging a crucifix above her bed and praying over the phone with her friend Donna Middleton. Together they'd read *The Purpose Driven Life,* a Christian self-help book. To Middleton, it was no coincidence that Roberts killed herself near a church. "Her whole focus was she's going to meet Jesus," Middleton said. "For Marianne to take a gun and shoot herself, it just had to be her ticket home."

In Middleton's use of the word "home" was a notion of Roberts going back to a familiar, welcoming place where she felt comfortable—a place that existed before WiFi and smartphones, perhaps. It seemed like Green Bank was initially going to be that home, until Roberts concluded that she didn't feel at peace in the Quiet Zone, either.

"I have realized that the only way out is to leave my physical body and return to the spirit world," Roberts wrote in a suicide note addressed "to anyone who will listen," which her sister shared with me.

> I know this decision will be difficult for people to comprehend, but it feels right to me, and I am not afraid. I am hoping that

> my death and this letter prompt discussions among elected officials, legislators, policy makers, parents, schools, and the general public about the dangers of wireless frequencies and that these discussions lead to laws that protect people from RFs [radio frequencies]. RFs affect everyone, not just those individuals who can feel them.

Many factors play into suicide. Several of the electrosensitives in Green Bank told me they'd had suicidal thoughts in their darker moments, though they were always pulled back by family or friends. One had even made an emergency call to a friend on a cellphone, with the device of her torment saving her in that moment of despair. Roberts almost seemed a victim of how isolated she'd become, of how quiet she'd made her life.

Stanga visited the forested spot where her sister's body was found. She tied two branches into a cross and propped it upright, the final marker for someone who had considered herself a kind of martyr to quiet. When I visited the area months later, nothing remained—though I did spot a rough-hewn wooden cross hanging from a garage several hundred feet away through the woods. I couldn't help but wonder if a neighbor had taken Roberts's cross for decoration.

I went and knocked on the home's door. A short woman answered. She said the cross on her garage was made by a nephew "when he was just a little fella." I asked if she was aware of the recent suicide nearby. She nodded. She believed she'd heard the shot but had simply thought it was someone "out sighting a gun." She'd discovered otherwise at church. "It happened right here in my own yard and I didn't even know it," she said. "It wasn't in the *Pocahontas Times*."

"Why do you think she killed herself here, of all places?" I asked.

"That's what my question is, *Why?* It's like that lady from Florida that came up with that dead child. *Why?*"

I nodded, though I had no idea what she was talking about. I later learned that, a year earlier, a few miles outside of Green Bank, someone was spotted dragging a corpse into the woods. Police detained the woman and discovered she'd been trying to dispose of her eleven-year-old daughter's body, which had injuries to the head and torso as well as signs of strangulation or suffocation. The woman had driven all the way from Florida after searching on her smartphone for "alligator ponds," "people killed on Virginia cliffs," and "Virginia highways that has huge cliffs." That apparently led her to the Quiet Zone.

THE SUNDAY AFTER CHRISTMAS. I scooted into a pew at Boyer Hill Mennonite Church, where Roberts had last parked. Maybe there had been something special about this church, and I hoped I might find some sense of closure or resolution by attending a service.

I sat next to a bearded man in a white dress shirt and American flag suspenders, among a dozen men on the right side of the sanctuary. A dozen women sat on the left, all wearing white bonnets and dark-colored dresses. The service started with a pimply teenager blowing into a pitch pipe to set the key. Then the entire congregation launched, a cappella, into harmony. It was beautiful. Their voices vibrated to the ceiling. The suspenders-wearing man sang in a voice so deep that our pew quaked.

It was the final Sunday of the year, and the pastor preached about crossing into a new year that held "12 new months, 52 new weeks, 365 new days, 8,760 new hours, 525,000 brand-new minutes, 31,536,000 new seconds." "Behold, I make all things new," he said, paraphrasing Revelation 21:5. The new year was a chance to start over, he said, a new start. I thought of Roberts, looking for a new life in Green Bank and beyond.

Roberts's final note made clear that she hoped her death would raise awareness about the dangers of technology. But she was not the redeeming figure she hoped to be. Each of her family members and friends spoke to me with a smartphone pressed to their ear. They still charged their devices by their bedsides. Nobody had changed their habits, despite Roberts's pleas. And if her own friends and family would not change, what hope could there be for the rest of us to reconsider our tech addictions? If death itself could not spur people to rethink the need for constant connectivity, what chance did the Green Bank Observatory have for getting the community's cooperation in preserving the Quiet Zone?

"We're so entrenched in technology," said Kohler, the coworker. "To turn back the clock is almost impossible."

FOR THE ELECTROSENSITIVES, the suicide was a warning: people flocking to Pocahontas County were discovering that the quietest place in America wasn't so quiet. They considered this a matter of life and death, and they wanted to bring attention to the issue.

But despite calls to the *Pocahontas Times,* Roberts's death never appeared in the county's sole newspaper. Historically, the media has not reported on suicides because of social stigmas and fear of inspiring copycats, though a growing consensus has emerged that, in certain cases, suicides can and should be covered in a nonsensationalistic, informative way that sheds light, sympathy, and understanding on mental health issues.

To editor in chief Jaynell Graham, however, this was nothing more than the case of a confused woman committing a senseless act of violence on herself. Not a story. "Anybody who kills herself because they can't find someplace they're supposed to be has a lot more going on than some kind of electromagnetic sensitivity," she told me.

Graham regularly got calls from sensitives around the country asking for information about Pocahontas, and she wasn't hesitant to tell them to stay where they were. A woman from Pittsburgh had recently called to subscribe to the *Times* because she planned to move to Pocahontas. "I told her it was the 'fallacy of elsewhere,'" Graham said. The grass is always greener on the other side, especially when the other side is called Green Bank.

Only once had the *Times* ever reported on the sensitives, with a 2009 profile of Schou. Since then, despite a steady stream of national and international media filing through Green Bank to report on the sensitive community, Graham avoided their story. "I think the problem in today's society is that we take abnormal behavior and try to make it normal," she told me.

When I met up with Graham over the winter holiday, I told her that I'd continued to look into the suicide case since she first told me about it at Thanksgiving. I'd learned that many people in the Green Bank area had heard about the death, yet nobody really knew what had happened, fueling rumors that her newspaper might have been able to correct. Graham sighed loudly, raised her eyebrows, and gave me an exasperated look. She said I was being "nosy."

"People have talked about their neighbors and gossiped forever," she said. "If I had to put one thing in the *Pocahontas Times,* I'd say, 'Don't believe a damn thing you hear and only half of what you see.'"

"But that's the whole point of having a newspaper," I said, frustrated by her approach to covering the news. My first job in journalism had been for a newspaper that was, like the *Pocahontas Times,* among the country's dwindling number of independently owned press outlets. I had met Jenna through the newspaper, where she worked as a reporter and editor. I thought it obvious that a robust press was vital to democracy, and that people should know the good and the bad happening in a community.

"What is it you think I should have put in the paper?" Graham asked.

Aside from Roberts's suicide, I said, there had been a recent, tragic death of a former member of the high school's championship forestry team. The twenty-two-year-old had been visiting her parents' home in Durbin when an out-of-control logging truck barreled down the road and fatally struck her in her own driveway. She'd gotten married only a month earlier. In memory of her, the high school's forestry club had added angel decals to its traveling van, then gone on to win the national championship.

"I got a call asking why it wasn't in the paper," Graham said of the death. "I said the police and insurance company are the only ones who need to know."

"What about holding the driver accountable?" I asked.

"Why would we pile on when he's going to have to live with this? Is it your business? Are you going to counsel him and help him through this? You want to throw the book at this guy? That's sensationalism . . . The difference between the *Pocahontas Times* and the *Charleston Gazette* is they don't know the people they're dealing with. Everybody knows most everybody here."

Seeking further to keep the peace, Graham had opted to not report that a former Marlinton councilman running for mayor had recently attacked someone at the local Family Dollar with a golf club. Graham said the former councilman hadn't formally launched his mayoral campaign, so he wasn't yet a public figure. But didn't assault in a public place automatically make someone a public figure?

The thing was, I'd heard about the golf club attack from Graham herself, just as I'd first learned about Roberts's suicide from her. News in Pocahontas quietly spread through chatter, not print.

Who was I to tell Graham what to cover? But the criticism was also coming from the community, with one person calling her paper "nothing more than a promotional tract."

"You know what?" Graham said to me. "We *are* a promotional tract. From an economic standpoint, we are a dying county, so you've got to support everything."

She had a point. Amid massive layoffs across the news industry, Pocahontas was lucky to have a newspaper at all. I didn't face the economic reality of figuring out how to keep the newspaper running.

In the vein of controlling the narrative, Graham focused on positive news. Suicide didn't fit into that picture. Nor did neo-Nazis, for that matter. She didn't consider her newspaper to be a watchdog. She once told me the Southern Poverty Law Center was "a bigger terrorist organization" than the National Alliance, apparently because of the nonprofit's dogged monitoring of hate groups. She later said she'd been joking. She thought the SPLC could be "overzealous," while the National Alliance, in her judgment, was never a threat to the community. It seemed to sum up her approach to the news. Live and let live. Stop being nosy. It walked a dangerously fine line between respecting people's privacy and being permissive of a hate group.

"It's hard to get news about Pocahontas County because it's not in the newspaper," I once told Graham.

"Everybody here knows what the hell's going on," she said. "*You* don't. But *you* don't live here."

"My impression is that a lot of people in Pocahontas wished they had a better sense of the news," I said.

"Well," she said, "tell them to start a fucking newspaper."

CHAPTER TWENTY-ONE

"Shutter the Place and Move On"

DOWN THE HILL FROM the National Alliance, a path of plywood planks and flattened cardboard led across a muddy yard to De Thompson's RV. I knocked. Thompson swung open the door with a big, toothless smile. He said he'd just been thinking about me—a scary notion. He welcomed me inside. It was late afternoon, a couple of days into 2019.

It was hard to imagine Thompson, his wife, and their two Rottweilers all crammed inside the RV. A futon faced two small metal chairs, each duct-taped with black foam for padding. *Die Hard* played on an old Magnavox television mounted between the front seats, where a black cat was curled in a ball. In the rear of the RV was a double bed and a tiny wash closet. In the kitchenette, family photos decorated the cupboard doors.

I took a seat on the futon beside Thompson, who was fiddling with a carton of cigarettes. A window looked toward the valley of Little Levels, with Droop Mountain in the distance. The setting sun was turning the sky orange and yellow, layering the rolling hills with hues of shadow. It was a sublime view of the Appalachian Mountains. Sure beat my apartment view in Brooklyn.

I told Thompson that I'd seen his name in the *Pocahontas Times.*

He laughed. Months earlier, police had made a marijuana bust near the National Alliance property, seizing a hundred plants. Seven of the plants—the "prettiest plants in the county"—were his, Thompson admitted. During the bust, police also found an unregistered gun in Thompson's car, which brought a charge for illegal possession of a firearm. Adding further to his troubles, he'd recently been robbed of $600 worth of pot. As if to check all the boxes for a hopeless situation, Thompson said he'd become addicted to opiates after accidentally taking speed laced with Subutex.

"It is a horrid, wicked drug," he said of Subutex, which is marketed for treating opioid dependence but is highly addictive itself. "I've come off heroin, OxyContin. No problem really. But this shit here, it's pretty wicked."

Thompson's smartphone rang. It was connected to satellite internet, rustic as the place was. Linda was calling to say she'd purchased pot from their supplier and would be home soon with their Rottweilers. "You're probably going to get pawed with a dirty paw, because it's muddy as fuck," he warned. Three adults, two dogs, and a cat sounded like a crowd inside the RV, so I got up to leave for the National Alliance compound.

"Ain't nothing going on up there," Thompson said. The National Alliance, he added, had recently sold off most of its land.

BOYD THOMPSON ROAD was a soupy, sludgy, impassable mess, forcing me to park a mile downhill from the compound. As I walked on from Thompson's RV, slipping in the mud, an ATV rumbled out of the woods. The driver waved and asked where I was going. I said to the National Alliance. He offered me a ride, which turned into a detour on his ATV trails.

"Want a beer?" he asked as we sped into the woods. I declined, but he still reached for a Michelob Ultra from a cooler, stuffing the can into a koozie and taking a swig. "Drive around in a car drinking beer and you get a DUI," he said over the growling motor. "Drive around in a side-by-side back in the woods drinking a beer, they ain't going to do nothing!"

His name was Derrick Miller. He lived in Charleston and came to Pocahontas on weekends "to cut loose." His father had purchased this twenty-five-acre property in the 1970s. His neighbors were never troublesome, he said, just annoying when the heavily tattooed National Alliance members rode around shirtless on four-wheelers.

"Just because they have beliefs don't mean I have to believe them," Miller said. "You do your thing and I'll do my thing. As long as we don't cross paths and have a problem, it's all right. But if we do cross paths . . ." He flipped open his jacket to reveal a gun holster with a .45 Colt revolver. "When you're out riding and stuff, you have to carry a pistol. You never know what you might find."

Miller dropped me off near the compound's entry gate. The sun had set, and the fast-dropping winter temperature was refreezing the mud. Ice crystals crunched under my feet.

THIS WAS MY FIRST TIME arriving in the dark, which made the compound all the creepier. Without my car, I felt even more isolated. I didn't know who I might find, except that it was likely to be an armed neo-Nazi. But I was due to leave Pocahontas the next day, so I didn't have time to come back.

The lights were off at the cottage that Hess had supposedly been turning into a "VIP headquarters." I walked on and rounded a bend to the main building that once held the business offices. A thin man stood outside smoking a cigarette. I called out a greeting and said I was looking for Jay Hess. He let me inside.

I saw a person sitting on a couch, then realized it was a mannequin wearing a National Alliance trucker hat—mannequins suddenly being a distracting theme. Hess popped out of a side room. He wore a heavy brown jacket, as it was chilly inside. His skis were propped in a corner. Rock music played over a stereo. It took him a moment to recognize me, but then he didn't seem surprised that I was back. He introduced me to the other man as "the journalist who went into the cave."

"I heard you sold some land?" I asked.

"We sold off some property up top of the mountain," Hess said, trying to downplay it. "The reasons are, basically, we're not using them and why pay taxes on property that you're not using?"

Actually, it was a major land deal. According to the assessor's office, the National Alliance had sold 267 acres in September 2018 for $215,600 to WV Coastal LLC, a real estate company run by the local realtor and property developer Oak Hall. Hall had been slowly acquiring surrounding land, and this purchase brought his holdings to about one thousand acres, including the land where De Thompson parked his RV. (Hall's parents, by coincidence, had been the real estate agents on Pierce's original land purchase in 1984.) The National Alliance and its nonprofit Cosmotheist Community Church still owned around seventy-five acres, including all the buildings, though Hess said they were also on the market. His role had changed from caretaker to real estate agent, as he was tasked with trying to sell the remaining land.

I realized that my role had also changed. Instead of monitoring the revival of one of America's preeminent hate groups, I was witnessing its last gasps in Pocahontas County. The National Alliance was planning to move entirely to Tennessee, near the home of the chairman.

Hess said he'd come to Pocahontas County to "honor" the memory of William Pierce. "I hated to see this place go under if I didn't at least make an effort," he said. "After one year, that's obviously not happening . . . Young people have to support themselves, and it's tough to get work here. The only people who will come out here are old people that are retired like me. I'm sixty-five and collecting Social Security."

I'd always been hard-pressed to find much local support for the National Alliance. Hess and David Pringle had declined to give me the names of their allies in the community, and the only Pocahontas-born resident who ever admitted to me that he'd been involved was a man named Shawn Kelly. He worked at a local prison and had run for county sheriff in 2012, placing third in the Democratic primary. I'd heard him referred to as "Nazi Shawn," owing to how he'd been spotted often at the compound over the decades and had once operated a nearby bar called the Eager Beaver, which came to be known as "the Nazi bar." His body was covered in provocative tattoos, which I first saw when he walked shirtless into McCoy's Market in Hillsboro. "Caligula" was tattooed across his chest, with a crowned skull on each of his pecs, all below an image of an AK-47 assault rifle. Six-foot-five and broad shouldered, he also had a tattoo of the date 3/23/1933, which was when Germany's government granted Hitler dictatorial powers. The names of John Wilkes Booth, Benito Mussolini, Augusto Pinochet, and Saloth Sar (Pol Pot) were scrawled across his skin. Amid it all was a tattoo of his wife's name and their wedding anniversary. When I asked why he didn't have a tattoo of William Pierce, he responded, "I only have so much skin." Kelly invited me to go skateboarding with him (not what I'd expected from a forty-year-old dad), and I watched as he sped at 35 mph down a stretch of the Highland Scenic Highway through the Cranberry Wilderness with a beer in hand. He told me he'd gotten involved in the National Alliance when he was in his teens. He'd spoken with

Pierce a number of times. ("Dude typed a lot," Kelly said.) In recent years he had occasionally gone four-wheeling with Pringle, but he didn't visit the compound anymore, and he dismissed any notion of the National Alliance having a local following.

According to Hess, the only people who showed up at the compound now were squatters and drug addicts. He said the recent marijuana raid had ensnared not just Thompson but people who had in recent months moved into the compound's abandoned buildings. A few had tried to build a meth lab. "That's why we're trying to keep the locals at arm's length," Hess said. "And since we can't get people from our organization to move up here, it's time to shutter the place and move on."

A book on the table caught my eye. The title was *Rising Out of Hatred: The Awakening of a Former White Nationalist*, by Eli Saslow of the *Washington Post.* Hess said he was reading the book because he was friends with the family of its subject, Derek Black. "I've known him since he was swaddled in diapers growing up," Hess said. Black's father was Don Black of *Stormfront.* Hess had lived near the Blacks in Florida and been a longtime contributor to the radio show. He proudly added that he'd personally come up with the name *Stormfront* in 1990, a detail that Don Black would later confirm. Hess had originally intended it as the name for his rock band.

Hess said he was enjoying the book because the author was "very objective, the way a journalist should be"—not what I expected him to say, given how the book traced Derek Black's disillusionment with white nationalism. The godson of David Duke, Derek was seen as the heir apparent to the white power movement until he disavowed the ideology in his twenties. In a November 2016 op-ed for the *New York Times* in which he renounced white identity politics, Derek called Donald Trump's racist rhetoric "destructive to the entire nation."

Perhaps seeking to change the subject, Hess suddenly asked, "Have you ever seen our secret underground bunker?"

My eyes lit up. For two years I'd been picking up rumors about secret tunnels and caves at the compound. I'd heard there was a massive bunker underneath a warehouse, and a pit so deep that light couldn't reach the bottom. Jerry Dale, the former sheriff, said an informant once told him of caves where Pierce created a water collection system that could provide a clean drinking supply should he need to retreat underground during the collapse of civilization that would lead to a race war from which he would emerge victorious (or so he envisioned). Mystery swirled in the community around the neo-Nazi camp, just as it did around Sugar Grove and Green Bank, with people passing rumors about a system of bunkers and tunnels under each property.

Hess stood and peeled back the carpet, revealing a trapdoor in the floor. His friend grabbed a screwdriver to wedge up the hatch. A hole dropped down about eight feet. Hess said I was welcome to go inside. I had a headlamp because I'd figured I'd be walking in the dark, so I strapped it on and climbed down a wooden ladder into the hole—perhaps a rash decision, as entering a dungeon at a remote neo-Nazi compound deep in Appalachia was an easy way to disappear.

At the bottom of the ladder, I stood on dirt and looked around. There was some fresh gravel scattered on the floor, as if the hole had been filled in at some point. The entire "bunker" was about the size of a closet. It wouldn't even function as a root cellar.

"It's nothing," I called up.

I heard Hess and his friend laughing.

"Well anyway, you got to see the only bunker that I know of," Hess said.

The entire organization was like that trapdoor to nowhere. The

National Alliance presented itself as a shadowy force with power and influence that reached throughout the community. In reality, the Alliance was more like this shallow pit, underwhelming and unimpressive. The neo-Nazis could certainly be dangerous. Their ideology was despicable and their aims criminal. But in many ways, they survived on an illusion of evil grandeur.

Hess said law enforcement had seen the bunker while inspecting the building after Pierce died. "They cleared out a bunch of chemicals from the laboratory upstairs," he added.

"There's a laboratory?" I said.

"Well, we might as well give you the full tour," Hess said.

I climbed out of the hole. Hess dropped the hatch and rolled down the carpet. I followed him upstairs. A door labeled "Laboratory" opened to a room filled with chemistry equipment and books, including a hardbound multivolume set of *Encyclopedia of Chemical Technology*. The shelves were lined with beakers, microscopes, glassware, capacity resistors, an autoclave, bottles of sinister-sounding chemicals, and bags of rock salt. It looked like an ideal place to make a bomb.

I asked why the National Alliance would need a chemistry lab. Hess mentioned how a member named Harry Robert McCorkill, who lived briefly at the compound, had earned his Ph.D. in chemistry from the University of Manitoba and briefly taught at Harvard University—as if that explained the point of having a chemistry lab here. Upon McCorkill's death in 2004, he bequeathed a $250,000 estate (including a collection of ancient coins) to the National Alliance. The Southern Poverty Law Center said the gift could have been a "lifesaving financial lifeline" for the National Alliance, but Canadian courts blocked the transfer on the grounds that McCorkill's beliefs ran counter to public policy.

I poked into another room labeled "Business Manager." On the floor was a pile of clunky keyboards, printers, scanners, and monitors, all more than a decade old. Balanced atop a pile of papers was a letter, still sealed, addressed to Pierce. I opened a random drawer of a tall filing cabinet, spotting folders labeled "Canadian Intelligence," "Tax Returns," and "Publishers." There were a pair of snowshoes, an amateur telescope, and paper star charts. Hess said he might auction it all off as an estate sale.

The other half of the upstairs once held the National Alliance's recording studio for its radio show. Hess had turned it into a living quarters for himself and his cats. An AV closet was packed with hundreds of video cassettes and DVDs, including of the sci-fi flick *The Thing* and the 1935 Nazi propaganda reel *Triumph of the Will.* The odd collection, like the organization itself, was somewhere between despicable and dopey.

Back downstairs, I spotted a Geiger counter on a table. It had been purchased decades earlier in case of a nuclear catastrophe, Hess explained. Pierce and the National Alliance had retreated to the mountains of Appalachia for reasons not too different from the U.S. government when it surveyed the area for bunker sites in the 1950s. "If there was some major crisis in the world, like nuclear war or some sort of catastrophe like an influenza outbreak, this would be a good place to get away from the chaos where a remnant could survive," Hess said. "People have been talking doomsday for the fifty years I've been involved in this movement. At some point . . ." His voice trailed off. He seemed tired of living for the apocalypse.

I later emailed Pringle to ask about the land sale. He was still in Nebraska managing a gun shop, though he'd run into controversy with the state chapter of the anti-fascist group Antifa, which had publicized where he worked. He said he opposed the property sale, but he acknowledged that it was near impossible to recruit people to live in Pocahontas. It was too far from a Walmart, he said.

"All the legacy NA members are now gone—the hive of activity, brotherhood, and attachment to The Land is now gone, too," Pringle wrote back. "I recently saw the NA referred to as a 'book club for old men.' When I read it I had to agree."

CHAPTER TWENTY-TWO

"Don't You Forget This"

IT WASN'T EASY CONVINCING JENNA to spend a two-week holiday in Green Bank. "I feel like you're stealing Christmas from me," she'd said when I proposed the idea, expressing particular concern about food. We couldn't expect an invitation to the Sheetses' table every night, and procuring fresh groceries in Green Bank was a challenge. The Sheetses, among others, took a two-hour drive every few months across the state line to Harrisonburg, stocking up on food from Costco and Sharp Shopper. So we packed a two-week supply of hardy fruits, vegetables, and coffee beans. Knowing we wouldn't have WiFi, we also packed several ethernet cords.

When we unlocked the door of our apartment at the observatory, however, we found a single internet jack, meaning all our extra ethernet cords were useless. I usually went alone to Green Bank, so I hadn't thought to bring an ethernet splitter that would allow multiple laptops to get online simultaneously. Within minutes, we were bickering over who needed internet more. Jenna was busy with her biggest work project of the year, and I was communicating with local people by email. I sped over to Trent's, where Bobby Ervine sold me what he thought might be an ethernet splitter. It was actually a landline phone splitter. The thing was practically an antique. I remembered the old adage "If Trent's doesn't have it, you don't need it," which in this case meant we were wrong in thinking it necessary for two people to be online simultaneously.

I'd also underestimated the challenge of finding a Christmas tree. I'd figured we'd pass a tree farm on the drive to Green Bank, which didn't happen. So I asked Bob Sheets where we might buy one. He seemed confused by the question. Buy a tree? Who in their right mind pays for a tree when surrounded by forest? Precut trees in New York City were selling for $25 a foot, making a typical eight-footer $200. Bob said he'd happily take our $200 if we wanted to stop by, borrow a saw, and cut down any tree on his hundred-acre property. He added that he wouldn't actually charge us.

After an hour of tromping through the Sheetses' woods, we realized that perfectly conical Christmas trees rarely grow in the wild. His property ran up to a barbed-wire fence with a wooden sign that read "U.S. Govt. Prop. No Hunting. N.R.A.O." On the other side, we spotted a shrub that could, with some imagination, pass as a miniature Christmas tree. So I hopped over the fence, cut down the shrub, and drove it back to our guest apartment. The gangly evergreen had about six branches, just enough to hold a dozen ornaments.

Our simple tree could have been an analogy for our time in Green Bank. Everything was pared down. There was one main road, one ice-cream brand at Trent's, and, at best, one activity every couple days, be it a trivia competition, live bluegrass, or a holiday party. Jenna and I went skiing once at Snowshoe, dined out once at the Elk River Inn, and attended a party at Jay Lockman's home, where he and his band played in a music circle. (I knew better than to try to join along this time.) Those few social events became more meaningful amid the quiet of our days, as if the quiet—the radio quiet, the audible quiet, the social quiet, all of it—enhanced our ability to listen, to hear, to appreciate companionship.

"There's something about being here that forces you to live simply," Jenna said, "and that creates headspace to want to read and write and do things I normally find taxing or a chore." Our New York City life was a daily bombardment of choices: Pizza or Thai? Whole Foods or Key Food? Bicycle or subway? R train or N? It was a paradox of choice, an anxiety-provoking number of decisions involved with almost every activity and task, with pervasive connectivity enabling a constant flow of alternate options and changes in plans, almost causing a kind of paralysis, like a child unable to make a decision in the cereal aisle. "You have to make so many choices to even make a choice," Jenna said of our "normal" life. Getting away from the bottomless options of Amazon and Google was freeing. It was a relief to be restricted to the basics.

Even such short visits to the Quiet Zone could be perspective altering, shocking outsiders like us into realizing, or perhaps just remembering, a less connected way of life. I once spoke to a young man named Greg Barber who was touring the observatory as a side trip to a ski vacation. His time in Pocahontas was causing him to reconsider a habit of sleeping with his smartphone. The Quiet Zone forced him "to live in the moment, like, *I'm here on planet Earth and this is where I'm at and this is my world right now,* versus in the digital world of where your friends are at, what information you want to look at, what game you want to play, what Facebook feed you want to be on." The quiet was a jolt, a shake of the shoulders to look up from the screen.

Just hearing about Green Bank could be powerful. When I told a friend in New York about the Quiet Zone, he locked his smartphone in his apartment for a month, using it only at his desk, just to prove to himself that he could go without it. Seemingly everyone whom I told about Green Bank responded, "I want to go there," because many people see the value of taking a break from tech. But because of the purposeful addictiveness of smartphones and social media, we often revert to default mode, which is to be always connected, always online. "People don't succumb to screens because they're lazy," as the computer scientist Cal Newport writes in *Digital Minimalism,* "but instead because billions of dollars have been invested to make this outcome inevitable." How many times had I vowed to take a break from the internet for a day, only to still find myself online?

There's a cliché that absence makes the heart grow fonder, but Jenna didn't find that she appreciated cell service more by being away from it, and I never reached a new fondness for WiFi after all my time in the Quiet Zone. Instead, we regained an appreciation for being a bit disconnected. Despite the fact that Jenna spent most days working on her laptop, we each found ourselves enjoying unprecedented stretches offline, unplugged. It was liberating.

It was also temporary. We were both too busy to feel bored or lonely. We never felt isolated, because we knew we could leave. And we didn't have to struggle with slow internet, because we could hook up to the observatory's fast connection. We streamed *Jeopardy!* and *Better Call Saul* every night with dinner. We were also helped by a crucial piece of hardware. A week into our stay, Rudy Marrujo, the tech-savvy HAM radio operator, loaned us an ethernet splitter. "I doubt you'd have been able to find one of these anywhere in West Virginia," he'd told me. Underscoring the challenge of getting *any* computer equipment, West Virginia is the only state in Appalachia without an Apple Store—not that an Apple Store would carry an old piece of technology like an ethernet splitter.

We never had to take on the real challenges of life in Pocahontas: the struggling economy, the food deserts, the rural internet. As much as I had come to appreciate the people, culture, and rolling hills of Pocahontas, there were downsides. Jenna, as a Korean native, acutely felt the area's lack of diversity, from the food to the prevalence of Confederate flags. And while the Quiet Zone forced residents to have a fundamentally healthier relationship to technology, even in Green Bank I found that I was one of the only people without a cellphone. I was still an outsider, still an anomaly.

AT A CAROL SERVICE on Christmas Eve, I'd felt like I knew half the congregation. Jenna and I sat a row behind the family of Laurel Dilley, the high school math teacher. They sat in the same row as Heather Niday, the radio station manager and wife of Chuck Niday, Quiet Zone cop. Behind us sat Diane Schou and her husband, Bert. Debbie Ervine from Trent's shared the piano bench with Pastor David Fuller's son, who tooted on a harmonica.

When the lights dimmed for a nativity reenactment, the church was silent and attentive. There were no ringing phones, no blinking screens, no temptations to scroll through social media or check email. Two children dressed as Mary and Joseph walked down the aisle and sat by a manger, soon joined by the three wise men, those foreign astrologers who had been on a mystical quest to find a savior. I thought of how so many people were on a quest to find quiet, which manifested itself in Green Bank, with the electrosensitives worshipping the giant white telescopes, finding healing from the sacred quiet. They had faith in Green Bank to save them from the ills of the modern world.

For the electrosensitives seeking relief from their pain, for the astronomers in need of a quiet sky, for the hippies desiring a peaceful landscape, for the tech-addicted tourists forced to go offline, the

Quiet Zone was an unexpected refuge. It was an escape, at its best, from *ourselves*. Perhaps quiet, in itself, could be a savior, redeeming us from our own noise. The electrosensitives were right: we do need a break from our devices. The hippies were right: we do need to reconnect with the land. The astronomers were right: we do need to be silent to listen.

Before returning to New York City, I said goodbye to Betty and Ebbie at Trent's. One of their stores, Trent's 3, had recently shut down and been bought out by the Par Mar gas station chain. The restaurant Station 2 was also up for sale, as if the businessman Buster Varner were giving up on the area. How long could Betty and Ebbie hold on to their way of life? With the National Science Foundation preparing to announce its decision on the observatory's future, who knew how much longer the facility would remain open?

"Y'all come back now," Betty always said when I left Trent's, as if she expected me to pop back in any day. She'd tease me to bring Jenna more often to Green Bank, making me feel like she was welcoming us both into the community.

"We'd like to be around her some," she said.

"Maybe she'll be my wife one of these days," I responded.

"Don't wait too long, she might decide to get somebody else. You better marry 'er and keep 'er."

"Maybe we should get married here."

"Well, that'd be great. The church would be a great place for you to get married."

"And you and Chuck can sing in the choir," I said.

"And Debbie can play the piano."

"And you can make mac and cheese."

"Don't you forget this," she said. "We've got it all worked out."

Betty waddled away from the register, her arthritic knee swinging.

CHAPTER TWENTY-THREE

"The Never-Ending Story"

"**YOU DIDN'T BRING YOUR GIRLFRIEND** this time?" Betty Mullenax asked. It was a warm summer day in 2019. I stood by the register inside Trent's General Store.

"She couldn't come," I said. "And she's no longer my girlfriend."

"Oh," Betty said apologetically.

"She's my wife," I said.

"Well, congratulations to you!" said Betty's daughter Debbie Ervine, who was standing nearby.

"And we're having a baby," I said.

"You're becoming a papa real quick, aren't you?" Betty said. "Well, congratulations. Maybe next time she'll be along."

"Maybe the whole family will be along."

It was good to see Betty. She'd had a tough summer. Her husband, Ebbie, had died in June. He'd gotten pneumonia and spent several weeks at West Virginia University Medicine's Ruby Memorial Hospital in Morgantown, three hours away. On the day he was discharged, he suffered a stroke while walking to their car and passed away soon after. The afternoon of his funeral was the first time Trent's ever closed on a day that wasn't a Sunday or a federal holiday. About 150 people attended the service, so many that Pastor

David Fuller delayed the start so everyone could squeeze inside. Betty had saved a handful of bulletins, and she retrieved one from the back room to give me.

"We were married for sixty-five years," Betty said. "See if you can make yours last that long."

"I'll be lucky if I live that long," I said. "I'd have to live to a hundred and one if I wanted to be married for sixty-five years."

"He was a good one," she said. Her lip quivered as she held back tears. She turned to the register to ring up another customer.

THAT SUMMER, the National Science Foundation announced it would continue to fund about two-thirds of the Green Bank Observatory's annual $14 million budget for another five years, which came as a huge relief to many astronomers, though not as a major surprise. It had become understood that the NSF's review was a hard nudge for the observatory to become more financially self-sufficient, and in that regard the initiative was successful.

As for whether the National Security Agency influenced the decision, the answer was befitting for a spy organization: it was a secret. Anthony Beasley, director of the National Radio Astronomy Observatory, told me that the impact to Sugar Grove from the Green Bank Observatory's potential closure had been discussed and it would have been "a cheap thing" for the NSA to reach out to the NSF on the observatory's behalf. As for whether that happened, "I don't know," Beasley said, "but I'm not sure that I would." The NSA did not respond to my requests for comment.

Aside from NSF funding, the other third of the observatory's budget was coming from tourism, from a collaboration with the North American Nanohertz Observatory for Gravitational Waves (NANOGrav), and from the Breakthrough Listen initiative to search for alien life—even if the search for E.T. seemed as whimsical as ever. Breakthrough Listen had recently announced that a survey of 1,327 nearby stars found no signatures of technologically capable life beyond Earth. But other discoveries were still in the offing. Over the summer, the Green Bank Telescope detected the most massive neutron star ever observed, with 2.17 times the mass of our sun crammed into a sphere only 18.6 miles across, contributing to an understanding of just how dense an object can get before imploding into a black hole. Such discoveries could continue until the NSF put the observatory back up for review, at which point the question would again arise: What is the Quiet Zone worth to us?

The observatory was already strategizing ways to bring greater appreciation to Green Bank. One idea was to host "digital detox" retreats, offering conference space to corporate groups that might value being "away from the distraction of connection," according to Michael Holstine, the business manager. He had also broached the idea with the NSF and the state historic preservation office of applying the Quiet Zone to the National Register of Historic Places. "We think it's a national treasure and don't want to see that changed," he told me. Green Bank already had one National Historic Landmark: the Reber Radio Telescope, built in 1937 in Illinois and relocated to the observatory's campus. The NSF's recent review of the property had concluded there were at least forty additional structures of historic significance. But while a telescope or building is a tangible object, humans have proven less adept at valuing something invisible and nebulous like radio quiet.

In just the few years that I'd been visiting, the Quiet Zone had changed irrevocably, with the amount of cell service beamed into Pocahontas essentially tripling. At the start of 2016, the only cell provider in the county was AT&T, which had antennas in Hillsboro, Marlinton, and atop Snowshoe Mountain Resort. Since then,

AT&T and T-Mobile had started transmitting from Caesar Mountain near Lobelia; T-Mobile had installed antennas in Buckeye and Marlinton; and T-Mobile and Verizon were now in discussions to provide cell service atop Snowshoe, joining AT&T on the ski slopes.

In the basement of the observatory's science offices, I found Chuck Niday amid a tangle of electronics, toolboxes, radio equipment, and machinery. I asked for his latest assessment of local WiFi. He swiveled to his computer and opened a map that showed all the hotspots he'd found recently: about 175 within two miles of the observatory and 355 within about a five-mile radius. At last count, the observatory estimated there were about 150 households within two miles. There now appeared to be more WiFi signals than homes, if that was even possible.

"This means everybody has WiFi," I said with astonishment.

"That's pretty much true," Niday said. "We go out on these site inspections, back on some godforsaken road, and anywhere there's a house, there's almost always a WiFi signal. And it's gonna get worse."

Interference was also coming from within the observatory. Niday had recently seen an employee pay for a meal at the cafeteria using a credit card app on a smartphone, a device that was supposed to be powered off at the observatory. "This is somebody that should have known better," Niday said.

In response to the problem of smartphones and WiFi, the observatory was considering building a tall wall or a large dirt berm around its property, a way of physically shielding the telescopes from radio noise. Since the community couldn't—or wouldn't—respect the quiet, a last option was to hide behind an actual wall.

But a wall wouldn't do anything about what was overhead. The astronomers were also voicing concern about proposals from private firms such as SpaceX and Amazon that wanted to roll out global

WiFi through a network of tens of thousands of orbiting satellites. The proposition was also alarming to the electrosensitives.

"That's a death sentence for me," Jennifer Wood said of global WiFi. To escape it, she had developed plans for a kind of "hobbit house" with thick clay walls and a sod roof that would block the incoming radio waves. She had also helped lead a protest in Washington, D.C., against global WiFi. "We need to start lobbying for radio quiet zones throughout the United States and the world," she told a small crowd from the steps to the U.S. Supreme Court.

Across the hallway from Niday, Paulette Woody sat at her desk, her hand in a brace. That summer, she'd undergone surgery for carpal tunnel in her right hand. If ever there was an indication of the rising amount of paperwork that the Quiet Zone administrator faced, it was that Woody's hands were failing under the load. In her first week back, Woody had processed "a couple hundred" coordination requests. Applications were still primarily for cell service, radio, emergency services, amateur radio, microwave, and television. She hadn't seen many 5G installations yet.

"Heaven only knows what's going to happen after 5G," Woody said. "It's going to be 6G, 10G, 12G, it'll be like the *Rocky* movies. It's the never-ending story of cell and technology."

Beyond threats to the Quiet Zone, Woody was anxious about what tech was doing to her own family. Her granddaughter had an insatiable appetite for YouTube videos and once used up Woody's monthly allotment of satellite internet data in two days.

"I remember when I was a kid, the social place was the kitchen table," Woody once told me. "That was the original social media, and we've lost that. Now when I go to visit with people, they're always strapped to that phone, even in the living room or around the kitchen table."

"There's something that works really well," she said. "It's called O-F-F. Just turn the thing off."

Or you could not have a cellphone, I suggested.

"You don't have a phone?" Woody responded, bewildered, when I told her I didn't own a cellphone. Even the Quiet Queen had a flip phone. "Oh my gosh, a millennial without a cellphone? You are living rural, aren't you? You are just an oddball."

QUIET WAS NO LONGER a default of living in Pocahontas. Cell service had also arrived to High Rocks, spurring the young women's camp to create new rules to help break youths off their devices. Campers were now asked to check in their smartphones for the duration of their stay, and this hard line against tech had turned into the organization's biggest recruiting hurdle, according to Sarah Riley. "The idea of stepping away from your phone is a bigger and bigger and bigger challenge that takes you far outside of your comfort zone," she told me.

Going offline had become way scarier than being within a mile of a neo-Nazi organization, or what was left of one. High Rocks was, by virtue of its quarter century existence and influence on the community's youths, a subtle rebuke to the legacy of William Pierce. High Rocks praised diversity, challenged youths to examine their biases, and sought to build inclusive communities. Thousands of young men and women—some of whom had family once associated with the National Alliance—had received support from High Rocks and were now carrying that mission forward.

Meanwhile, the National Alliance was still trying to sell its remaining land. It's a "million-dollar property," the organization's chairman, Will Williams, told me. He thought a church group might like to turn the half dozen buildings into a religious retreat, which sounded far-fetched. David Pringle told me that "most of that place needs to be burned down at this point." Aside from the property's disturbing history, any buyer would have to expect the occasional neo-Nazi or white supremacist to wander through. Williams said the National Alliance retained a right-of-way to the top of the mountain so people could pay their respects where Pierce's ashes were scattered.

It was probably in Williams's best interest that he not visit Pocahontas County much anymore, since De Thompson told me he wanted to kill him. At some point, Williams had blocked the road to the compound, which enraged Thompson. He'd gotten his rifle, snuck up a nearby ridge, and trained his gunsight on Williams. He said he would have pulled the trigger had Jay Hess and two young men not walked into the scope.

"I don't talk about me putting crosshairs on people, but now that I'm closer to dying I don't give a fuck," Thompson said when I stopped by his RV that summer. "The only reason I haven't done something too outrageous is I've got to pull something off to make enough money to leave Linda and my dogs in a little better condition than what they are."

I didn't doubt that Thompson could be dangerous, especially as he launched into a bizarre, bloody tale of having "lived through the Pablo Escobar days down south," working as a kind of hatchet man for criminal drug organizations in Florida and South Carolina. He said he'd robbed a string of pharmacies and homes, with a resident once waking up as Thompson tried to steal a diamond ring from the man's finger. "Last thing he saw was the butt of that .357," Thompson said. "I went *wham! Wham! Wham! Wham!* 'Go to sleep, motherfucker!'"

"That's a violent assault," I said, somewhat shocked.

"That's the only time it ever turned violent," Thompson said.

"Nobody ever knew I was there. They'd just wake up and everything would be gone."

Linda's voice crackled over a handheld radio. Through the static, I heard her say she'd found a half dozen ginseng plants in the forest. According to state law, ginseng season was still two weeks away, meaning Linda faced a fine of up to $1,000 for illegal foraging. But the Thompsons were desperate for cash. Ginseng was going for around $500 a pound. And I was getting roped into their scheme for selling it.

"Keep digging money," Thompson replied to Linda. He invited me to have a seat in a plastic lawn chair. He said he'd reached an agreement with the new property owner, Oak Hall, to continue parking his RV on the land. I asked what kind of "terms" they'd reached. "I've known the boy since before he was a squirt out of his daddy's dick, okay?" Thompson said. "It's that simple." In short, he wasn't budging. (Hall confirmed that he didn't "have any intent to remove" the Thompsons.)

Thompson started quoting from "Desiderata," a 1927 prose poem, which he said he reread every couple years "when things get bad." He considered it his personal code. "Go placidly amid the noise and haste, and remember what peace there may be in silence," he recited. He paraphrased the rest of the poem.

"There's only one thing I can't agree with," he said. "'Speak your truth quietly and clearly.' I can speak my truth, but it won't be fucking quietly."

Linda showed up with a sack of ginseng slung over a shoulder. She was using two golf clubs as hiking poles. Missing all her teeth, her mouth appeared even more sunken in than when I'd last seen her. She looked sickly thin, just bones holding her gaunt frame upright.

"I've been pulling weight for everyone," she said, referring to Thompson and his two sons from a previous marriage. "It's just been really stressful." Adding to the difficulties, Thompson had recently totaled their van, and their Subaru had died. Without a vehicle, Linda had no way to sell her illicit ginseng.

"Do you need a ride into town?" I offered.

"How much time do you have?" she asked. She had a ginseng buyer in Renick, a forty-five-minute drive south into Greenbrier County, but she'd first need to make a call from the area's only spot with cell service, in Hillsboro.

"You do have a cellphone that works, correct?" Linda asked.

I shook my head.

"You don't have a fucking cellphone?" Thompson said. Even he had a smartphone, just no data plan because money was tight. "*Really?*"

AROUND THE COUNTY, the tourism bureau had begun distributing a new brochure that read:

> Life is fast paced. But vacation shouldn't be. Welcome to the National Radio Quiet Zone. 13,000 square miles of land, federally protected from artificial radio wave interference, where the secrets of the universe can be revealed by the world's largest steerable radio telescope at the Green Bank Observatory. Meaning no cell service. No WiFi. Just you, your family, and our grand outdoors. Find your peace.

The brochure seemed to signal a change from when tourism director Cara Rose told me the Quiet Zone was a hindrance to attracting visitors. Was radio quiet now seen as an asset to tourism?

Rose corrected me. The brochure was meant to explain the limited WiFi and cell service, not to promote the quiet. Quiet would never be a selling point in and of itself, she said. Tourists came for skiing, mountain biking, hunting, fishing, and camping. "We won't market Pocahontas as a destination to come because of the Quiet Zone," Rose said flatly.

I told her I'd check back in a few years to see if things had changed and there was a greater appreciation for the Quiet Zone.

"We will *never* put in an ad 'No cellphone coverage,'" Rose said. "*Never.* That will *never* happen."

The brochure was inaccurate, in any case. I'd first seen it at a restaurant called Dean's Den that advertised WiFi on its front door. Andrew Dean, formerly the chef at Mountain Quest Inn and the son of its owners—the people who believed the Quiet Zone contained a portal to another dimension—ran the establishment, which happened to be in the same building in which the neo-Nazi Craig Cobb had once operated Gray's Store, Aryan Autographs and 14 Words, LLC. The store now sold organic crackers, fancy cheese and sausages, and craft beers. Dean told me that about 10 percent of his customers were electrosensitive. Some asked him to turn off the WiFi, but he declined because the internet attracted other patrons—and anyway, the WiFi didn't stop the electrosensitives from coming. The day I stopped in, the special was a chicken masala curry made with chanterelle mushrooms that Bob and Elaine Sheets had foraged from their woods.

BOB SHEETS CRACKED OPEN three bottles of beer in celebration. He and Elaine had just shared the news that they were expecting another grandchild, so I'd told them that Jenna and I were also having a baby. We clinked our bottles with a "cheers." A new generation would have to determine the value of quiet and if it was worth preserving.

I looked out over the yard and past a wooden fence to their son Jed's house a quarter mile away, close enough for convenient grandparenting. I knew the Sheetses didn't have WiFi, which Bob touted as a point of pride. His mother had worked at the observatory for three decades; he'd been the facility's longtime neighbor and public champion; he regularly poked fun at the outsiders who went into withdrawal without cell service. He had a reputation to uphold. But I'd never asked if his son's home had WiFi.

"Yeah, he does," Bob admitted. If their son had WiFi, then any potential harm to the Green Bank Telescope was likely already being done, regardless of what the Sheetses did.

"So why don't you have WiFi?" I asked.

"I would feel hypocritical," he said. "We function. I can sit down at the computer and do whatever I need to do."

"You're one of the last houses in the area without WiFi," I said.

Bob nodded. He named a few homes that he thought didn't have WiFi, but he agreed they were the exception, not the rule, even if the media continued to portray an alternate reality about Green Bank. The *New York Times* had recently published a story titled "The Land Where the Internet Ends," which claimed that Green Bank "residents do without not only cellphones but also Wi-Fi, microwave ovens and any other devices that generate electromagnetic signals." Of course, most residents did have cellphones, WiFi, microwave ovens, and myriad other gadgets. The following year, another *New York Times* feature would describe Green Bank as a place "where Wi-Fi is both unavailable and banned." Which was news to everyone around Green Bank.

The way residents responded to most media articles about the Quiet Zone reminded me of how people from Appalachia responded to *Hillbilly Elegy,* the bestselling 2016 memoir by Yale Law School graduate J. D. Vance about escaping poverty and dysfunction in

Appalachia. The memoir had inspired a book-length rebuke titled *What You Are Getting Wrong About Appalachia* as well as an essay collection, *Appalachian Reckoning: A Region Responds to* Hillbilly Elegy, whose opening essay noted that for nearly 150 years "the region has been incessantly 'discovered' and then 'rediscovered' by a long series of novelists, journalists, social scientists, satirists, and documentarians, most—if not all—inspired by the irony of Appalachian Otherness. How can a region defined by the Euro-American frontier myth be so different, so *far behind,* the perceived American mainstream?" Didn't that sum up so many portrayals of Green Bank? How could this place be so different, so "backward," as to not have cell service or WiFi?

It was a hot August afternoon, and I savored the cold beer as it bubbled down my throat. I'd first sat on this porch more than two years earlier, when I was welcomed to the Sheetses' backyard party despite nobody knowing who I was. The Sheetses had shared so much with me, even hosted me and Jenna for Thanksgiving, asking for nothing in return except one thing.

"I read something the other day that I think applies to us," Bob said. "A journalist was interviewing a guy in the coal fields. The miner said, 'I'll talk to you as long as you don't do "poverty porn." Don't hold us up for the world to look at as if our poverty is entertaining.'"

In the same way that the miner didn't want to be portrayed as the face of white poverty, Bob didn't want to be a caricature for digital disconnection. The yarn about Green Bank as "the place where the internet ends" felt like another kind of media exploitation, repeated ad nauseam for an outside world. "It's 'disconnectivity porn,'" Bob said. He didn't live in a "dead zone." And he didn't need to be brought to life with "connectivity." Far from it. He was already living.

EPILOGUE

"Masters of Social Distancing"

IN THE WEEKS LEADING UP to our child's birth, I was often asked if I was anxiously checking my phone in case Jenna went into labor, almost as if the questioner were projecting their digital anxiety onto me. When I replied that I didn't own a cellphone, I was typically met with a look of horror. "But what if she goes into labor and can't get in touch with you?" the person would ask in an accusing tone, as if I were breaking an implicit civil code and should be dragged before a jury to account for my disconnection.

Jenna ended up having a scheduled induction. We took the subway to the hospital, and the next morning the two of us became three. Using her smartphone, I snapped a photo of us in the delivery room, the baby still with streaks of blood in his hair, and shared it with family and friends. Jenna's mother, visiting from Korea, recorded video on her iPad. The devices were valuable in those moments. If I'd learned something from Green Bank's complicated relationship with technology, it was to be pragmatic in my own usage.

For Jenna's maternity leave, we relocated to a fixer-upper cottage in an area of rural Connecticut known, appropriately enough, as the Quiet Corner. My parents lived nearby, ready to help with baby care. As we settled into a routine, I began to hear the voices of the electrosensitives in my head. Some research suggested that smartphone radiation possibly posed a higher risk to children, owing to their smaller heads and thinner skulls. So we took some precautions. Jenna's smartphone and my iPod stayed distant from the baby and outside the bedroom where we all slept. We got an analog clock for the bedside. We opted out of an array of "smart" baby care options, like the diaper that texts when it detects poop or the socks that track heart rate and blood oxygen levels.

Initially, ours was a temporary move to the countryside. Then COVID-19 swept the nation, and our relocation became indefinite. At first, quarantining had the feel of a slumber party or prolonged snow day. My parents had all the time they wanted with their newest grandchild. Jenna perfected her pizza-making skills. I built a patio and vegetable garden. We watched a robin build a nest in our pergola, lay four eggs, and deliver grub to her brood until they were big enough to fly off. We weren't alone in embracing a low-key lifestyle. The initial months of the pandemic turned into the longest-recorded period of audible human quiet in history, based on data collected from 268 seismic monitoring stations around the world. Highways emptied. Stores shuttered. Construction stalled.

Meanwhile, the digital noise was blaring. Screens became a necessity for education, employment, and social connections. The online overload was even felt around Green Bank. After schools went remote, Pocahontas County High School math teacher Laurel Dilley started getting migraines from toggling between her computer screen, iPhone, and iPad all day as she tried to coordinate with students on email, Facebook, and Instagram. It was the first time she felt compelled to set digital boundaries in her life, and by late spring she'd begun purposefully turning off her devices from 4:00 P.M. to 7:00 P.M. to give herself a break.

On the flip side, Dilley felt lucky to be connected at all. She estimated that at least half her students lacked reliable internet, making remote education nearly impossible for them. Ruth Bland, the technology coordinator for the county's schools, asked the U.S. Department of Homeland Security to bring in two temporary mobile internet providers, called cells on wheels (COWS), to increase internet availability—with the understanding that the COWS would be far enough from the observatory that they wouldn't interfere with the telescopes. Homeland Security denied her request because of the area's low population, underscoring a catch-22: the small population meant companies invested little in internet infrastructure, which meant residents had slow internet, which meant Bland needed to ask for emergency help, which was denied because of the low population. The best Bland could do was open WiFi access at all public schools and libraries, with the exception of in Green Bank. She also asked teachers to stagger their Zoom sessions so they wouldn't overload the rural internet lines.

At Mathias Solliday's home, the internet was so slow that he couldn't download slide presentations, submit assignments, or watch the video lectures necessary to remotely finish his freshman year at West Virginia University. Every day, he drove to a friend's house ten minutes away, where he parked in the driveway and tethered to the WiFi, which came from a faster internet provider. His younger brother sometimes joined him, the two sitting side by side in the back seat of their parents' Highlander with their laptops open to schoolwork.

Still, Solliday was glad to be in Green Bank during the pandemic, if he had to be anywhere. For a couple days, West Virginia had the distinction of being the last U.S. state without a case of COVID-19, making the remoteness of Pocahontas especially enticing to people fleeing cities. License plates from New York to Florida began appearing in the parking lot of Trent's General Store as outsiders stocked up and packed into the county's campgrounds. "It was completely nuts," Donnie Ervine told me. "People we didn't know were trying to buy twenty pounds of hamburger at a time." Shelves emptied. All the while, Ervine's ninety-two-year-old grandmother Betty Mullenax kept working the register, refusing to wear a face mask on account of some "flu bug."

People were also panic-buying land in Pocahontas, with home sales surging about 30 percent for Red Oak Realty, according to owner Oak Hall. "The number of people making knee-jerk decisions is through the roof," Hall told me. It was the biggest land boom in a decade. The observatory closed to tours, but visitors could still walk around the telescopes. Bob Sheets, who biked regularly around the observatory, was seeing as many as thirty people on the site daily, and they repeatedly asked him if he knew of any properties for sale.

To limit the influx of out-of-staters, the governor shut down all campgrounds in late March 2020. Some visitors resorted to sleeping in their cars. Sheriff Jeff Barlow repeatedly found a woman with electromagnetic hypersensitivity overnighting at random parking lots.

The county reported its first case of COVID-19 in April, though some residents believed the virus had already been spreading undetected through the community. Hanna Sizemore, for one, was bedridden for two weeks in March with symptoms that a physician said clearly indicated COVID-19, even if her test was negative. She suspected it was a false negative. She had asthma, and she battled coronavirus long-haul symptoms through the rest of the year: fatigue, shortness of breath, extreme swings in heart rate, and the sensation of having a heart attack. But despite the distance from robust medical care, the Sizemores didn't feel it was a hardship to live in Pocahontas during the pandemic. Quarantining almost came naturally. They had fifty pounds of beef in their freezer, plus a pantry stocked with flour, sugar, and yeast. Once a week, a neighbor delivered a dozen fresh eggs to their doorstep, a kind of rural version of FreshDirect; the Sizemores paid via PayPal.

"We are the masters of social distancing," said Ruth Bland. "I have a neighbor that lives a football field away, and my other neighbor is a half mile away. You have to cross a river and a field to get to anybody." Allen Johnson said it was already normal for him and his wife to spend weeks in isolation; he only ever purchased toilet paper twice a year, and his pantry, root cellar, and freezer were full of food before the pandemic hit. Same for Bob Sheets, who had two freezers full of venison, beef, and other food. He also had fifty-five gallons of fresh maple syrup from that spring's tap.

The electrosensitives were also relieved to be in Pocahontas, though for slightly different reasons. Several told me that COVID-19 was linked to cell towers and 5G cell service (a baseless conspiracy theory), which was why the virus was worse in cities. To them, the Quiet Zone even had the power to ward off a pandemic.

MY SON SITS on my lap, munching on Cheerios that I've arranged in a neat row along my desk to distract him from banging on my laptop. As I attempt to type this paragraph, he reaches for my face, then pushes against the desk with his feet. He pulls a tube of ChapStick from a drawer and, with the utmost intent, drops it to the ground, a game that he repeats with my wallet and credit cards.

Isolated with an infant during the pandemic, I found myself struggling more than ever to strike a balance between being physically present and digitally connected, often failing at both. Always home, I couldn't be with my family without seeing my laptop and thinking about work. It was also impossible to immerse myself in work without seeing or hearing my son. There was no separating work from play from family time. It all melted together.

I was not alone in feeling digital burnout, and I hope there's a silver lining—that the pandemic might push the widespread tech fatigue to a critical mass that spurs a cultural shift, or at least a cultural nudge. After feeling trapped behind screens through so much of the pandemic, perhaps when we emerge from the gloom we will choose to be physically present with others, setting aside our devices for a spell.

People have urged me to get a phone for my son's sake. I think it's the opposite. For his sake, I will keep some distance from the device, in part because a smartphone would be another distraction from him. It would also undermine the kind of example I'd like to set—of someone who is attentive, able to sustain a conversation, able to stay in the moment. As it is, I don't live up to that goal. I'm far from the most attentive dad. It'd be that much worse if an iPhone were also competing for my attention.

I realize my position comes with entitlement. For many people, circumstances demand they be constantly connected for work, for education, for childcare. That doesn't undermine my stance. Rather, it creates a greater onus on me to show that a cellphone-free life is still possible. As Jenny Odell writes in *How to Do Nothing,* the attention economy's grip over many people makes it "even more important for anyone who *does* have a margin—even the tiniest one—to put it to use in opening up margins further down the line." So I will continue to stick out my elbows.

My son may well become the most hyper-connected person I know. That's fine. It's his choice. I just want it to stay *a choice* and not a requisite for living.

DO I REALLY have to say it? I also hope my biracial son never experiences the discrimination that so many minorities have faced. Through the course of this book, I was forced to reckon with the very real terror of white nationalism in the United States and the insidious legacy of William Pierce, which was on public display in January 2021 when a mob stormed the Capitol. Under the banner of Trump flags and white nationalist emblems, protesters erected a makeshift gallows and called for legislators to get "the rope." Analysts saw clear allusions to a scene in *The Turner Diaries* where "traitors" to white America are hanged from tree limbs and electric poles. As photos circulated of rioters breaking into the Capitol, it seemed surprisingly easy to throw democracy into chaos, which was perhaps the event's ultimate achievement. In *The Turner Diaries,* a mortar attack on Congress is valuable to insurrectionists not for its death toll, but for its "psychological impact" in showing that "not one of them is beyond our reach."

In the wake of the raid, Amazon stopped selling *The Turner Diaries*. But Pierce's writings were beyond the tech giant's control. The book has seen a renaissance online as Pierce grows in stature as a visionary for the Far Right movement, according to Heidi Beirich, the hate groups analyst. Today, people can become radicalized without formally joining an organization, making brick-and-mortar organizations like the National Alliance unnecessary. White nationalism no longer requires real estate. What Pierce wrought in Pocahontas has metastasized into a nebulous, leaderless, online echo chamber, even as forest reclaims his mountain compound. "Dr. Pierce's name and his writings are still readily shared all over the internet," said Billy Roper, who now leads a white nationalist prepper organization in Arkansas. "Since he died, his presence has only grown."

Sarah Riley of High Rocks once told me that white supremacy "is a scary thing whether it's two miles away from you or five hundred miles away." I'd responded that it certainly seemed scarier when it was in your own community.

But she was right. White nationalism was never just a Pocahontas County problem. It was always *our* problem. It was always my problem, even if it was as invisible as the radio noise polluting Green Bank.

Acknowledgments

THE PEOPLE OF POCAHONTAS COUNTY made this book possible, and I owe them my deepest gratitude. They gave me lodging. They gave me food. They gave me their stories. Everyone named in this book, and many more unnamed, has my thanks.

Bob and Elaine Sheets provided guidance and nourishment. Betty Mullenax let me hover by her register at Trent's while I bugged her for hours. Jaynell Graham was generous with her time and always willing to entertain my questions, even if she didn't like where they were leading. Tony Byrd and Becky Sheets were among many families who extended their table to me. Bob Martin sat for many beers and many more tales. Bob and Ginger Must trusted me, and for that I am honored. Thanks to Jerry Dale, David Jonese, and Jeff Barlow for explaining the challenges of law enforcement in the Quiet Zone. I thank De and Linda Thompson for helping me understand another side of the county, for taking me into the depths of the earth, and for showing me how to forage ramps. Among the area's many inspiring teachers and educators, I want to thank Sarah Riley, Laurel Dilley, Greg Morgan, Ira Brown, Kristi Tritapoe, and Joanna Burt-Kinderman for speaking with me. I also thank Allen Johnson, a tireless and courageous champion for Appalachia.

At the Green Bank Observatory, thanks to director Karen O'Neil for allowing me to wander her facility's hallways pestering her employees. A huge thanks to Mike Holstine for his patience with my many questions and for once letting me borrow a dongle so I could get on wired internet. Thanks to Jay Lockman for his ability to dumb down radio astronomy without making me feel dumb. Paulette Woody gave me parenting advice; Carla Beaudet taught me the difference between a ferret and a ferrite; and Chuck Niday demonstrated how to be a smooth radio jockey. Bob Anderson, Galen Watts, Wesley Sizemore, Hanna and Dane Sizemore, and so many more people were gracious in explaining their work. I am impressed with the entire staff and their efforts to expand human understanding of the cosmos.

At the National Radio Astronomy Observatory in Charlottesville, thanks to Anthony Beasley, Kenneth Kellermann, and archivist Ellen Bouton, who helped me sift through folders in the archives and enthusiastically hunted down answers to my questions. At the Robert C. Byrd Center for Congressional History and Education at Shepherd University, archivist Jody Brumage also dug up dusty boxes that provided illuminating details.

I want to thank Diane Schou, Sue Howard, and others in Green Bank's electrosensitive community who spoke with me. For their sake as well as for astronomy, I hope the National Radio Quiet Zone continues to be recognized as a valuable resource. Thanks also to the family and friends of Marianne Roberts.

I would also like to acknowledge the help of David Pringle, as much as I disagree with what he stands for. He spoke with me for many, many hours and openly shared about his role in the white nationalist movement.

Thanks to David Helfand for his encouragement, and to two additional faculty at Columbia University who provided crucial support. First, Terry Thompson accepted me into the Knight-Bagehot Fellowship at Columbia. Once there, Professor Samuel G. Freedman begrudgingly allowed me to take a coveted seat in his book-writing seminar—begrudgingly, because my initial book idea was so half-baked. He showed no mercy in critiquing my writing, which was a great gift. For years afterward, his steadfast enthusiasm for *The Quiet Zone* pushed me onward, even when my own morale was flagging.

Thank you to my literary agents, Larry Weissman and Sascha Alper, who took a chance on an unpublished author. They were excited from the start to learn more about Green Bank.

Thanks to Jessica Sindler, my first editor at Dey Street Books, who had an incredible ability to see the narrative thread through my bloated drafts. She brought a big-picture vision to this project, which included lining up the talented cartographer Mike Hall to create a beautiful map. Hilary McClellen, fact-checker extraordinaire, saved me from potential embarrassment. Thanks also to editor Nick Amphlett, who deftly led the manuscript across the finish line. I appreciated his patience and even-keeled nature through my many last-minute changes.

My aunt Kathleen Brown and my good friend Nate Paluck suffered through early drafts and provided generous feedback. They and other friends and family endured many rants from me about the Quiet Zone and the publishing process.

My parents have always been a source of unwavering support, and their faith in me is surely why I had the confidence to embark on this endeavor. They also provided a lot of child care, which allowed me time to work.

When I first drove to West Virginia with Jenna, we had only been dating for several months, and she could hardly have guessed what she was signing up for. I thank her for being a constant supporter, yet critical; for encouraging me to make extra calls, while dragging me out of rabbit holes; for reading more drafts than she would care to remember, yet always being willing to read another. Her patience with this project is underscored by the fact that we got married and had two children before the book was published. She's never asked me to get a cellphone, and for that I am also grateful.

Author's Note

THE BULK OF THIS BOOK was reported from in-person interviews, but I also relied on many media articles, journals, and books.

For the history of the Green Bank Observatory, I benefited immensely from *But It Was Fun: The First Forty Years of Radio Astronomy at Green Bank* (2007) by Felix J. Lockman et al. Lockman's Great Courses series on radio astronomy (2017) was a wonderful backgrounder on the field. I also drew information from *Open Skies: The National Radio Astronomy Observatory and Its Impact on US Radio Astronomy* (2020) by Kenneth I. Kellermann et al.

History of Pocahontas County, West Virginia 1981, published by the Pocahontas County Historical Society, provided an exhaustive chronicle of the area. The *Pocahontas Times* archive was also an invaluable resource. John O'Brien's memoir *At Home in the Heart of Appalachia* (2001) helped me better understand the culture. I gleaned insights from *Life Without Mercy: Jake Beard, Joseph Paul Franklin, and the Rainbow Murders* (2014) and *The Third Rainbow Girl: The Long Life of a Double Murder in Appalachia* (2020). Henry Rauch's research for the *Journal of Spelean History* (2018) revealed much about the strange case of Peter Hauer. Helen Zuman's memoir *Mating in Captivity* (2018) brought me inside the Zendik Farm.

On William Pierce, the National Alliance, and the white supremacist movement in America, I am indebted to Leonard Zeskind's *Blood and Politics* (2009), Mel Ayton's *Dark Soul of the South* (2011), and

Richard A. Serrano's *One of Ours* (1998). The Southern Poverty Law Center's online archives were a constant reference. Further context into the modern white nationalist movement came from *Everything You Love Will Burn* (2018) by Vegas Tenold, *Bring the War Home* (2018) by Kathleen Belew, and *Rising Out of Hatred* (2018) by Eli Saslow. I also drew details from Robert S. Griffin's controversial *The Fame of a Dead Man's Deeds* (2001) and Kelvin Pierce's memoir *Sins of My Father* (2020), which was the source of the FBI field reports.

On tech addiction and cellphones, sources included Nicholas Carr's *The Shallows* (2010) and *The Glass Cage* (2014), Sherry Turkle's *Alone Together* (2011), Adam Alter's *Irresistible* (2017), Jean Twenge's *iGen* (2017), and Cal Newport's *Digital Minimalism* (2019).

Index

A MAP OF

POCAHONTAS COUNTY
WEST VIRGINIA

10 miles

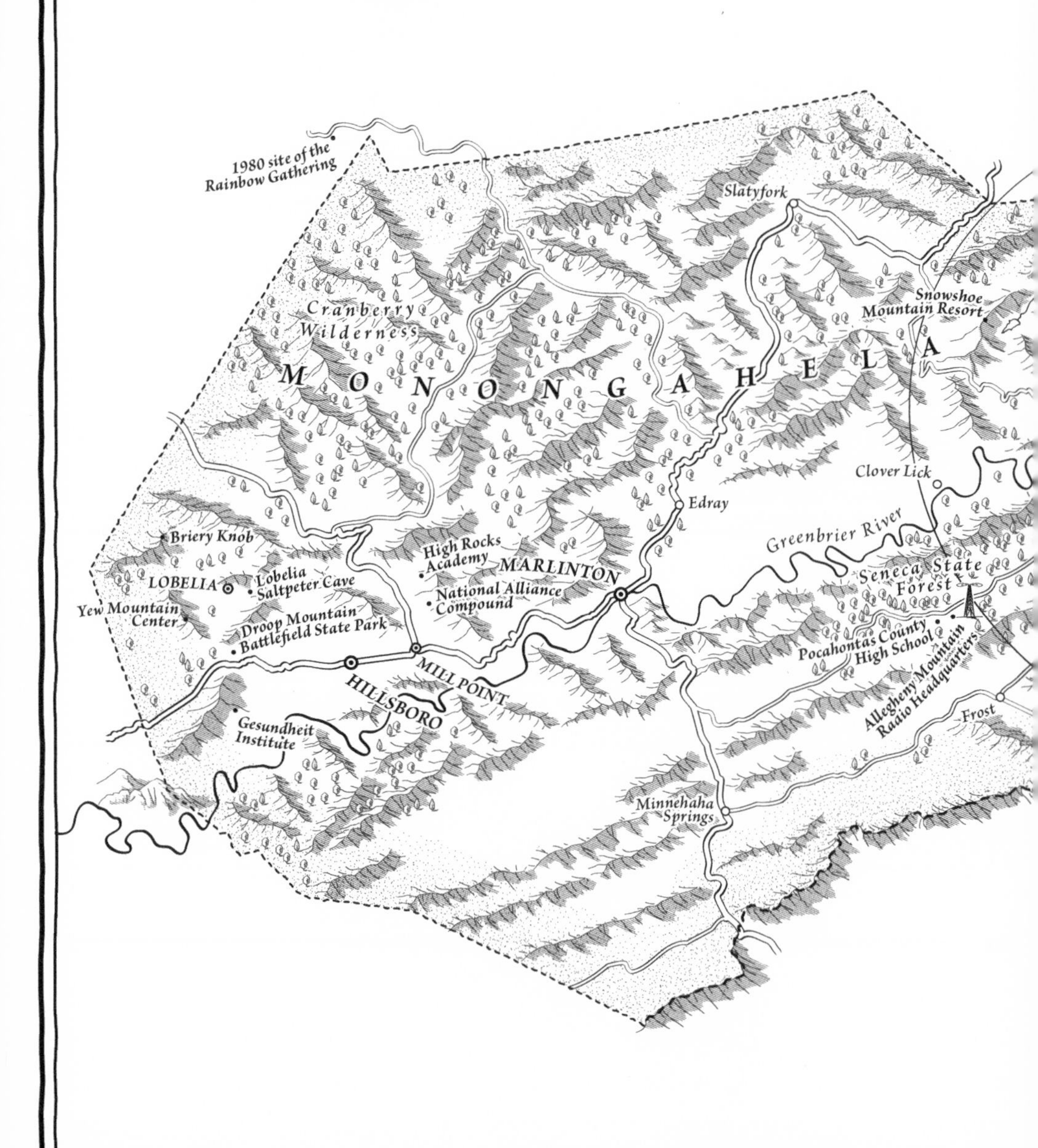

Shavers Fork
N A T I O N A L F O R E S T
Cheat Mountain
DURBIN
Bartow
Spruce Knob
Boyer Hill
Mennonite Church
Trent's
General Store
ARBOVALE
Observatory
Elementary-Middle School
Public Library
GREEN BANK
N
10-mile radius
100 miles
PITTSBURGH
PENNSYLVANIA
OHIO
COLUMBUS
MARYLAND
WEST VIRGINIA
WASHINGTON, D.C.
POCAHONTAS
COUNTY
Sugar Grove
Station
Green Bank
Observatory
CHARLESTON
KENTUCKY
CHARLOTTESVILLE
WHITE SULPHUR
SPRINGS
National Radio
Quiet Zone (NRQZ)
RICHMOND
VIRGINIA